A Guide to Effective Publishing in Astronomy

Coordinated by: *Claude Bertout, Chris Biemesderfer and Agnès Henri*

edp sciences

17, avenue du Hoggar
Parc d'activité de Courtabœuf, BP 112
91944 Les Ulis Cedex A, France

Contents

Foreword

This collection is one of the outcomes of a workshop for journal authors and referees, held originally at the XXVIII General Assembly of the IAU in Beijing, in August 2012. The workshop was organized by the editors and publishers of the major research journals published in the field of astronomy. While the workshop itself was conceived anew, the written materials relied on the book that was compiled for the *Scientific Writing for Young Astronomers* (SWYA) school in 2008, by the editors of *Astronomy & Astrophysics*. The current volume contains a number of articles from the SWYA[1] book, several of which have been updated, as well as several new articles. The topics that are covered include how to write a good paper, how to be an effective reviewer, and how the modern scholarly journal system works. The intended audiences are young astronomers, although the material presented here will be beneficial for astronomers of all ages.

Astronomers today rely on a wide array of digital services and resources as they conduct their research, especially as they write articles about their research for the scholarly record. The research journals have existed for a fairly long time. However, many of those resources didn't exist as recently as twenty years ago, such as the ADS or SIMBAD or the virtual observatory. Every indication is that the research enterprise is enriched by the introduction of re-usable digital assets and the ability to link them together on modern digital networks. Today's modern journal platforms are evolving so that the formal literature – the scholarly record – is well-integrated into the online environment to support the research needs of scientists.

The journals are just one class of digital resource at your disposal as a scientist, and it is very important that you use them effectively as you report your own research. Formal communication of results in the scholarly literature is a fundamental responsibility for all scientists, and writing articles for the journals is a skill that requires attention, and a little practice. An understanding of the environment in which the journals operate should help new authors remain oriented as they gain experience. Hopefully, the essays in this volume will help guide young astronomers as they develop their formal writing talents.

It is a rewarding time to be working as a scientist. There are many excellent tools at our disposal. Learn to use them well.

Chris Biemesderfer
American Astronomical Society

---

[1]SWYA: Scientific Writing for Young Astronomers, *EAS Publications Series* **49** (2011).

A Guide to Effective Publishing in Astronomy
© The authors, published by EDP Sciences, 2012

A BRIEF OVERVIEW OF THE PUBLICATION PROCESS IN ASTRONOMY

Claude Bertout[1]

Abstract. We review the *modus operandi* of the main research journals dedicated to publishing the results of astronomical and astrophysical research with the aim of helping young researchers who are getting ready to publish their first papers. We first provide a brief description of the publishing landscape in astronomy and continue with a discussion of the ethical requirements common to all peer-reviewed journals. We then explain how the evaluation of a paper proceeds after it has been submitted to one of the major astronomy journals, with a focus on the respective roles of editors, referees, and authors. We round up this overview with a short discussion of the topical issue of open access to scientific publications.

1 Introduction

You are a young researcher in astronomy, working hard on your PhD thesis, and your adviser just told you that the data you have acquired and reduced during countless observing nights, or the difficult theoretical computations you have worked so long on, represent sufficiently original and interesting material to "write a paper".

You are thrilled.

But the more you think about it, the more worried you become. Of course, you know that publishing papers in prestigious journals and giving talks in international conferences are the two main ways to present your results to the community of workers who share the same research interests. You also realize that the number and quality of your publications are crucial for obtaining a position in research. However you have never written a research paper before, and you have heard numerous tales of wicked editors whose main goal in life seems to be to stop your colleagues and friends from publishing their beautiful results. So you are understandably nervous about that first paper.

---

[1] A&A Managing Editor, Observatoire de Paris, 61 Av. de l'Observatoire, 75014 Paris, France

Relax.

In the following, we give you an overview of the publishing process in astronomy and tell you how to avoid the various pitfalls that could slow down the publication of your work or even (heaven forbid!) result in its rejection. We start in Section 2 with a brief, non-exhaustive description of the various astronomy publications, so you can decide which journal is most likely to publish your results, depending on your research field. We then review in Section 3 the ethical requirements common to all peer-reviewed astronomy and astrophysics journals, and continue in Section 4 with an overview of editorial practices, covering the respective role of editors, referees, and authors. Finally, we mention what happens after your paper has been accepted (Sect. 4.4), and briefly discuss in Section 5 the timely and thorny issue of open access to scientific results.

2 The Astronomy Publishing Landscape

The publication landscape in astronomy and astrophysics is much less crowded than in many other scientific fields: more than 90% of the original research is published in only four large international journals. These are the Astrophysical Journal (ApJ) with its offsprings the Astrophysical Journal Letters (ApJL) and the Astrophysical Journal Supplements (ApJS), the Astronomical Journal (AJ), the Monthly Notices of the Royal Astronomical Society (MNRAS), and Astronomy & Astrophysics (A&A). Together with the Publications of the Astronomical Society of the Pacific (PASP), these journals constitute what librarians call the core astronomy journals, needed by all active researchers in astronomy and astrophysics.

Along with these journals, there are several smaller international and national publications, some of them more specialized than others. A non-exhaustive list of international journals published in English includes Astronomical Notes, which accepts papers in general astronomy and astronomical instrumentation; Icarus and Planetary & Space Science, which specialize in planetary research; and Classical & Quantum Gravity and the Journal of Cosmology & Astroparticle Physics, whose intended audiences are clear from their titles. Several of these journals also publish invited reviews and conference proceedings. Examples of national journals published in English are the Publications of the Astronomical Society of Japan, Revista Mexicana Astronomia y Astrofisica, and Acta Astronomica. Some publications in national languages, *e.g.*, Astrophyzika, offer an English translation.

Besides research journals, there are publications devoted to publishing review material. As a young researcher, you are not likely to be asked to contribute to these, but you will find their articles invaluable for getting acquainted with the current research in your discipline area. The Annual Review publishes series that are each dedicated to a specific branch of science, and the Annual Review of Astronomy and Astrophysics and the Annual Review of Earth and Planetary Sciences are the best known review journals in astronomy and planetary sciences. The Astronomy & Astrophysics Review and Space Science Reviews also serve our field.

Nature and Science occupy a special place in the astronomy publishing landscape. While they are broad scientific journals whose range includes such wide fields as biology and physics, they sometimes publish accounts of major astronomical discoveries. On one hand, these two journals publish less than 10% of the articles that are submitted to them, so a paper in Nature or Science reflects positively on its authors. On the other, these communications are meant to reach a broad audience, so they are restricted to highlighting results of general interest.

In the following, we focus mainly on publishing in one of the core research journals, ApJ, AJ, MNRAS, PASP, and A&A, since it is likely that you will wish to publish your first research paper in one of these. These journals are published in English but accept papers from all over the world. They are all very wide in range, with subtle differences that are explained in the presentations by their respective Editors-in-Chief. For example, ApJ and AJ do not publish papers on instrumental developments in astronomy, while MNRAS, A&A, and PASP do. Papers submitted to these four journals should present new astronomical or astrophysical results or ideas of sufficient interest to the community as concisely as possible, while being comprehensive enough to allow for the duplication of the results by others.

The alert reader will immediately ask: *"How do I know that a result is of sufficient interest to the community?"* This is a valid but complex question that comes back regularly in discussions among editors, and the information content required of new submissions is something that varies with time. Fifty years ago, when astronomical information was much scarcer than now, it was easy for researchers to publish at least one paper after every short observational run on a 1 to 2 m-class telescope, because there were indeed new and useful ideas to be gained even from limited observations of exotic celestial objects. Today, the situation is entirely different, thanks to the huge amount of astronomical information that flows from large optical telescopes and radio arrays, space observatories, and surveying telescopes. As a consequence, the typical information content of a paper must be substantially greater for the journals and readers to keep up with the information flow and stay up-to-date with the science that follows from it.

This brings us to a related issue. Even for journals of wide purview, there are topics that are on the borderline of the subjects covered by astronomy publications. Examples include the theory of strange stars and some theoretical aspects of celestial mechanics.

To know whether a subject is covered in one of the major astronomical journals, the prospective author can evidently look at the journal's contents and search for similar articles. But there is an even easier way for an author to know whether an astronomy journal is likely to consider his work for publication: look at the reference list at the end of your paper. If none, or hardly any, of the cited papers have appeared in major astronomical journals, then it is likely that the editor will decide that the paper should be submitted to one of the specialized journals to which the manuscript refers. These editorial decisions are made on a case-by-case basis, as seen by the example of strange stars. Such a manuscript can be published in a wide-range journal if, *e.g.*, predictions are made that can be tested astronomically, say by the properties of pulsars, but will normally be rejected if it does not discuss astrophysical consequences. Similarly, a paper devoted to a theoretical issue in celestial mechanics that is not applied to at least one astronomical object will usually be rejected.

If a paper is not deemed acceptable by the major journals because it is too specialized, it doesn't mean that it can't be published. There are a number of highly regarded publications that are likely to accept it. To come back to the examples given above, an outlet for works dealing with the theory of strange stars could be Physical Review D, while Celestial Mechanics and Dynamical Astronomy is a natural place for theoretical works in that field. We mentioned earlier that MNRAS and A&A publish papers on astronomical instrumentation if these describe spectacular advances on instruments that are widely available to the community. They will not accept, for example, descriptions of more standard instruments or of observing facilities reserved to limited teams. These could, however, be published in other respected peer-reviewed journals that accept instrument descriptions.

3   The Ethical Requirements of Astronomy Journals

All refereed journals require their authors to adhere to the strict ethical principles that govern all academic endeavors.

Respect of intellectual property is the most important of these principles. The copyright on publications represents one way to protect intellectual property, so that academic journals usually request that authors transfer the copyright of their works to them. The transfer of copyright does not bar authors from advertising their work in the most efficient way (*e.g.*, by distributing it widely via abstracts and preprint/reprint databases). When speaking of copyright we are touching on much wider issues such as the so-called open access to scientific information. We come back to these in Section 5. We only need to stress here that the respect of intellectual property imposes several practical requirements on authors, and we mention them in turn below.

Plagiarism is usually defined as *"the act of reproducing text or other content from works written by others without giving proper credit to the source of that content"*. Citing a copied text literally is not the only condition for determining plagiarism, which also includes any paraphrased text that discusses an already published idea without citing its source[1].

Plagiarism is a major ethical breach, and it may also constitute a legal breach of copyright if the reproduced material has already been published. This is

---

[1] Reproducing or adapting in one's writings material taken from Internet pages, *e.g.*, from Wikipedia articles, without mentioning the original source also constitutes plagiarism. That Wikipedia is a free encyclopedia to which everyone can contribute does not mean that its contents can be used freely. Wikipedia's terms of use state clearly that *"Each copy or modified version that you distribute must include a licensing notice stating that the work is released under the Creative Commons Attribution/Share-Alike License 3.0 (Unported) and either (a) a hyperlink or URL to the text of the license or (b) a copy of the license.* Similar restrictions hold for most material available on the Internet.

particularly true when authors cite text from their own previously published works. Editors refer to this as *"self-plagiarism"*. Authors who wish to quote directly from other published work must cite the original reference and include any cited text in quotation marks, and figures may only be reproduced with permission. Because the research journals focus on publishing original research results, authors are discouraged from using any direct quotations and figures[2] of previously published papers. Software tools are available to help editors locate plagiarized material in submitted articles and are used routinely by journals.

Papers should cite previously published papers that are directly relevant to the results being presented or discussed. *Improper attribution – i.e.*, the deliberate refusal to cite prior, corroborating, or contradicting results – represents an ethical breach comparable to plagiarism.

Plagiarism, self-plagiarism, and improper attribution can result in the summary rejection of a manuscript. In the severest cases of plagiarism, offending authors can be banned from publishing for a determined period of time. In such cases, the Editor-in-Chief can also inform the Editors-in-Chief of the other professional astronomy journals of this ethical misconduct. Fortunately, such breaches remain rare, but it occurs often enough for us to mention it here.

When you submit an article, you attest that your coauthors have read the work and agree with the paper's contents. Adding prominent scientists to the authorship even though they had nothing to do with the paper is a practice that seems to be developing in some countries, perhaps in the hope that the article will be accepted more easily. This is, of course, another reason for summary rejection.

We should also mention that a given paper cannot be submitted simultaneously to more than one refereed journal. We must emphasize in this context that most journals do not systematically refuse papers that have been previously rejected by another journal. However, the editors expect a frank attitude from authors. If an author tells the editor about that paper's previous history, it is likely that it will be given a second chance to be published. Some authors choose not to tell editors when their submitted paper has already been refused elsewhere; but since all leading journals use the same pool of referees, chances are that the editor in charge of the paper will find out anyway. In this case, summary rejection is likely to follow.

Finally, a given work cannot be published in more than one refereed journal. Doing so, and we have seen a couple of cases in the recent past, is *not* the best idea for increasing the length of one's publication list. Instead, it constitutes the ultimate ethical and copyright breach, and can result in a long-term ban from any publishing.

---

[2]Figures that have appeared elsewhere (*i.e.*, copyrighted pictures) are usually not acceptable in articles unless they are absolutely necessary for understanding the article. In such a case, the author should request written permission of reproduction to both the original publisher of the figure and its author and send these documents to the editorial office when submitting the article.

4   The Peer-Review Process

Once your first paper is finalized and you have chosen a journal to publish it, all you have to do is to submit it for publication. This is done electronically via the web site of the journal. You will receive a confirmation that all the files have been received, and then your paper will be passed to an editor who will handle its peer review. The journals now rely on relatively large teams of scientific editors to cope with submissions in the various areas of astronomy and astrophysics. The peer-review process of submitted articles and Letters to the Editor is done in several steps described briefly in the following[3]. Steve Shore's chapter in this volume provides more detailed information on the peer-review process as seen from the editorial side.

4.1   First Step: Editor's Initial Reading

The editors in astronomy are responsible for the contents of their journal on behalf of their publication board. This responsibility can only be shared partially with referees since it involves issues both of journal policy (editors enforcing decisions made by the publication board) and of scientific excellence, for which the editors rely mainly on the expertise of referees. Deciding whether a submitted paper should be sent to a referee for a scientific evaluation is therefore the first question that the editors ask themselves upon receipt of a new article or Letter to the Editor.

4.1.1   Articles

In their initial reading of a new submission, the editors are not judging the scientific value of the paper in detail (this is the referee's role), but must gauge its potential scientific interest and information content, because both are important criteria for new submissions. A paper presenting an important result of high scientific interest can see its information content diluted by details or digressions that are unnecessary for understanding the result. Conversely, a catalog with large information content might have little *apparent* scientific interest but will be extremely useful to the community for further astrophysical research.

The large majority of all the articles that are received by journals pass this first test and are sent on to referees. The discussion of Section 2 explains why an editor might either not do so or might request some changes in the paper prior to sending it to a reviewer.

4.1.2   Letters to the Editor

In addition to the criteria described above, Letters should present novel results that require rapid publication. This criterion is not very well defined, though,

---

[3]This section is an adapted and abridged version of a previously published editorial by C. Bertout & P. Schneider (2004, A&A, 420, E1).

and its meaning has changed with time. There are now other ways to quickly communicate essential information than in journals. In particular, we note a clear shift towards using the *astro-ph* preprint server as a communication channel for timely results in the extragalactic community. In addition, publication times in the Main Journal are not much longer than for the Letters. What, then, makes a manuscript a Letter?

To qualify for a Letter, a manuscript should be both topical and of very high quality. In addition, the manuscript should usually not exceed four journal pages. Automatically, then, manuscripts are not considered for publication as a Letter if their contents appear less than well-timed – for example, if the reference list contains no paper on the subject from the past few years – or if it is too long. The length restriction has been relaxed at times, but a manuscript with five pages must be exceptional, and referees are given particular instructions about this.

In contrast, it is against editorial policy to sacrifice clarity to brevity. Trying to shorten a manuscript to Letter length at the cost of legibility is therefore not acceptable. It is also against editorial policy to defer essential information to a later, longer manuscript. An expert reader must be able to judge a scientific paper based on the information presented in the work or already available in the literature. A reference to work "in preparation" is not useful for evaluating a paper's contents, and the manuscript may either need expansion – and thus most likely exceed the page limit for a Letter – or need to be postponed until the article referred to becomes publicly available.

4.1.3 Some Style Issues

Editors may momentarily refuse to consider a paper not for scientific reasons, but because its style and language are obstacles to understanding the science. While linguistic and stylistic improvements can be made at the editorial office upon acceptance of the paper, the editors are not in a position to rewrite badly written papers before sending them on to referees, nor should they be. While non-native English speakers cannot be expected to write perfect English, those who have not yet developed full control of the language are urged to find a native English-speaking colleague or friend to read and correct their work.

A journal's instructions for authors provide precise guidelines concerning the style that editors expect, and new contributors are kindly urged to read them carefully before writing their first papers. The manuscript preparation as a LATEX file, now standard in all major journals, has put an additional burden on authors, but leaves much less room for mistakes during the final processing of the paper at the publishers. To exercise this control, however, the style guidelines provided in the instructions need to be followed exactly.

A few words about the length of papers are now in order. While the scientific content of many papers fully justifies their length, some submissions appear verbose and not concise enough. This is more often true of young researchers who use their thesis material for writing their first research papers. Thesis manuscripts are justifiably very detailed, because their authors must show that they have mastered all details of their own work and of the previous work on the same topic by others. In research papers, however, the work's background can often be reduced to the minimum compatible with proper referencing of previous work (see Sect. 3), and many details of the data reduction procedures or lengthy derivations of equations can be skipped, since the reader is expected to know the current state of the art of the research being discussed. Authors also need to cultivate the usual unadorned style of international scientific communications (cf. L.J. Adams's chapter on English language editing at A&A).

4.2   Scientific Evaluation by the Referee

Once the editor concludes that a newly submitted paper falls within the scope of the journal and is reasonably well organized, a referee is chosen to review and evaluate the scientific contents of the paper. Editors of astronomy journals look for the best possible referee for any given paper without resorting to pre-existing lists. Instead, they use ADS to find a potential referee within the group of people who have worked on the topic of the submitted paper in the recent past. If the field of research and/or the people involved are unfamiliar to the editor, she or he looks at the recently published works to find out whether there are competing groups on the same topic, who the most cited people are, and so on. With the online tools available nowadays and some experience, the editor can quickly get a rough idea of any astronomical research field, which is usually sufficient to make a good choice of referee.

It is important to note that because astronomical research exists in a framework of global cooperation, the peer-review process also needs to be global. This ensures that the scientific criteria for paper acceptance are comparable at the main astronomy journals, as demonstrated by their comparable rejection rates.

Most journals usually use only one referee for any given paper, except when a paper deals with more than one subfield of astronomy, in which case editors may request the assistance of a specialist in each subfield. This is, however, rare. We also usually request a second referee opinion in case of conflict between author and first referee (cf. Sect. 4.3).

In some cases, finding a potential referee turns out to be difficult. Examples here are papers in a highly specialized field. Another example are papers authored by a large group in a community, such as happens when new facilities go into operation; at this point, it is difficult to find an outside person competent enough to judge the technical aspects of a manuscript. In such cases, we sometimes ask the authors to provide a list of a few competent, independent potential referees, and most of them are quite willing to come up with a list. After a quick check with ADS, appropriate referees are then identified.

Referees focus primarily on the scientific contents of the paper. Journals provide guidelines to the various aspects of the evaluation that the editors would like the referee to address. A report prepared by a seasoned referee will not necessarily address all these questions explicitly but does so implicitly instead.

Once the editor receives the referee report, she carefully reads it and evaluates its usefulness and constructiveness before sending it to the author by going back to the paper and comparing the author's and referee's respective points of views and arguments. It is at this stage that the rare biased or offensive reports can be detected and edited[4], if needed, before they are passed to the corresponding author, who will share the report with her coauthors. The authors are then requested to revise the manuscript according to the referee report. Of course, editors expect authors to take into account all concerns expressed by the referee and editor and to provide a courteous and detailed reply.

After twelve years of experience as Editor-in-Chief, I can testify that the peer-review process works extremely well in the vast majority of cases. Many colleagues consider that refereeing is an integral part of their professional duties and are keen to deliver constructive and thoughtful reports within a reasonable time frame. Authors revise their works in accordance with the referee's requests and suggestions. In most cases, two iterations with the referee are needed before the referee recommends acceptance of a submitted manuscript. Sometimes, however, complications develop.

4.3   Complications in the Peer-Review Process

There are several possible causes of complications in the course of refereeing that we discuss below in some detail. When major problems arise, the Editor-in-Chief is likely to intervene in the process and help the various parties involved find a fair and reasonable solution.

Besides the ethical issues that may result in summary rejection from submitted papers (*cf.* Sect. 3), other less serious but more frequent problems can occur during the peer review. They are usually related to communication delays between the different actors of the scientific validation that result in delaying the decision of whether to publish the submitted papers.

4.3.1   Delay in Finding a Referee

Most senior astronomers are asked by the various astronomy journals to provide at least one referee review every year per journal, and the most well-known of them are asked to provide several reviews in any single year. This strong demand on referee time has the obvious consequence that referees may refuse to review a paper on the grounds either of being already busy with a paper for another journal or of having done several reports recently. However, we have noticed that referees seldom refuse to review a paper when they really want to know what it contains,

---

[4]A brief aside to correct a frequent misunderstanding: editors are perfectly entitled to edit referee reports before sending them to authors when their tone is condescending or offensive. They are also entitled to remove some referee requests that appear unfounded or to add their own suggestions for revisions that will improve the paper. The editors, not the referees, are responsible for their journals' contents. Referees recommend publication – or not – but the decision to publish is taken by the editors based on the referee reports and their own judgment.

so this appears a primary incentive for a scientist to accept a refereeing request. Initially providing the paper's abstract also helps referees decide whether they can competently deal with the paper. Nevertheless, one senses that the fraction of refereeing requests that are refused is slowly increasing over the years. One reason for this is the growing pressure on the time of senior scientists. In addition to the demand to referee journal articles, senior researchers are faced with an increasing number of reviewing and report requests from funding and evaluation agencies.

In many cases, the editor will quickly think of someone who would be eager to see the submitted paper so there will be no need to ask several people in turn, a process that usually results in long delays. In recent years, the median time between the time of submission registration and the time at which a regular article goes to the referee has been about a week. However, the tail of the distribution extends to more than thirty days. For the Letters, these times are somewhat shorter; nevertheless, there are exceptional cases where it takes more than three weeks before a referee has been located who is willing to review a manuscript.

4.3.2　Delay in Getting the Referee Report

As already mentioned, competent referees are often extremely busy with a variety of other duties besides their own research and/or teaching. The likelihood that the review request will end up being postponed or even forgotten by the referee is therefore not negligible. Editors usually send a first reminder to the referee three to four weeks (two weeks for Letters) after the task was accepted, and kindly request an answer. The worst-case scenario is when a referee answers the first reminder by saying that the report is under way, but after two or three reminders, does not bother to answer anymore and does not send a report. The editor then needs to start the process over and find another referee. This is when the longest delays occur. There is not much that the editor can do to reduce this delay, except to explain the situation to the second referee and kindly request a report on a very short timescale. In such cases, the editor will often contact trusted personal friends and colleagues who are kept "in reserve" for dealing with these emergencies.

Often, editors will be contacted by authors who believe that the referee is actually trying to delay the paper because of supposed competition with the authors. We would like to emphasize here that, in our experience, biased referees are extremely rare, much less common in fact than suspected by authors. In most cases where a conflict of interest can occur and the editor is not in a position to know that, the referee usually declines to review the paper and explains the reason frankly. In the past several years of practice at A&A, editors have witnessed only a few instances of obviously biased reports or delay that were the result of competition.

4.3.3　Offensive Attitude of Referee or Author

A problem that we experience more often with referees is a slight tendency to condescension. Whether there are grounds for such an attitude or not, it is understandably upsetting to the authors; therefore, the editors often edit reports that they perceive to be patronizing. In the few cases that border on offense, the editors do not send the report to the author but ask the referee to rewrite the report or look instead for a second, independent opinion.

Authors can also have offensive attitudes toward referees, in which case the editors request amended replies or edit them. Again, this is a rare occurrence. A more frequent attitude on the part of the authors is to say in their reply to the report that they have taken the referee's comments into consideration, whereas the revised text is basically unchanged. This is very upsetting to the referee and can quickly lead to a stalemate in the peer-review process. To avoid this problem as much as possible, the editors routinely request a detailed reply from the author to the referee report, including a description of the modifications to the text that were made as a result of the referee's remarks.

After this review of potential problems, we emphasize again that, in most cases, the peer-review process improves articles. According to an A&A author survey organized a few years back, 28% of authors find the referee reports very useful in improving the submitted articles, 52% find them useful, and an additional 18% find them somewhat useful. These opinions justify the role of independent peer review as organized by academic journals.

4.4   After Acceptance

After a paper is accepted for publication, it undergoes a final check by the Editor-in-Chief and is often sent out for language editing. English is a second language for many authors, and homogenization of the various English idioms used by authors is thus necessary. One chapter of this book deals specifically with language editing, and we refer the reader to this chapter for details.

Once this step is over, the accepted manuscript is electronically passed to the publisher for publication. Agnès Henri, who is in charge of the physics and astronomy department at EDP Sciences, the publisher of A&A, reflects on the complex role of the publisher in scientific publishing in a subsequent chapter of this book, so we refer the reader to her contribution.

One should perhaps clarify here the respective roles of language editors and of copy editors. The difference between their functions is straightforward: language editors make certain that the syntax of the article is grammatically correct and consistent and that the author's writing style in English conveys the scientific message clearly and concisely, while copy editors check that the final article complies with the journal's style sheet.

The copy editors' work often goes unnoticed by authors although it is absolutely essential for maintaining a professional quality of publication across all published articles. But who wants to deal with the fact that S/N, SNR, FWHM are in Roman characters unless they are accompanied by a specific value, in which case they must be written in italics; that there is a period at the end of a footnote (except when it ends with an internet link) but not at the end of the author's address; that Figure 3 is always abridged as Figure 3 unless it is the first word of a sentence? There are countless such rules that lead to a high level of consistency in the finished articles' style. Authors usually do not know about them, and do not have to because copy editors will work behind the scene to make sure that all articles look the same as far as style is concerned.

After an article is processed by the copy editors, it is scheduled for publication. It first appears online on the site of the publisher and the author is then encouraged to send a copy of the final, copy-edited and typeset PDF article file to arXiV[5] for archival purposes, while the abstract and other article metadata are sent to the NASA/Smithsonian Astrophysical Data Service (ADS) and other abstract repositories by the publisher. In parallel, you will possibly be asked, depending on the nature of your work, to send the original data published in your article to a data repository such as the Strasbourg data center CDS, which is contracted by some of the journals to archive the data discussed in the articles and make them available to the community. Authors are sometimes reluctant to share their data with others and argue that they are the owners of their data because they proposed and ran the observations that led to their acquisition. This is obviously mistaken, since these astronomers acquired the data on behalf of their financing institutions and observatories. These will eventually want to make all their data public to optimize the return on the huge investments needed to build and run observing facilities. Thus, the next years will be marked by an increasingly strong demand from the financing institutions to make all data used in articles available to the community, and the journals are likely to expand their collaboration with data centers with the aim to archive reduced data that have been validated by the peer-review process.

A few days after all of this is done, your article will appear on the website of the publisher. A few weeks later it will appear in print. Meanwhile, you will get congratulations from your colleagues for an interesting contribution and a job well done. It is a good occasion to have a sip of champagne before starting to work on your second research paper!

5   Toward Open Access for Astronomical Journals?

The economics of publishing is not necessarily something that interests astronomy students, but you are likely to hear discussions about open access to scientific publishing in your institute's cafeteria, so reading on might help you grasp what is at stake.

Publishing a scientific journal has a cost that someone must bear. Although the referees may be working for free (which means that the reviewing activity is considered to be part of their professional duties supported by the institution that pays their salaries), the costs of peer review includes the setup and maintenance of a secure article database and manuscript management system, the time spent

---

[5]Astronomy journals do not forbid authors from posting early, non-refereed versions of their papers on arXiv and recommend that they post (or re-post) there the final version of their article.

by the editors, and the salaries of the editorial assistants and language editors. Beyond that, the costs related to the production of the journal include the copy-editors salaries, the production tracking software, the preparation of the online and printed editions, the journal web site and archive, and the printing.

How do the major research journals in astronomy cover these costs? Interestingly, their economic models are very different (see the core journal presentations in the next chapter). These different models have consequences for the topical issue of open access.

The term "open access" (OA) has been a buzzword in the scientific publishing world for several years now. What is it exactly? As the online encyclopedia Wikipedia[6] notes, the OA concept comes in a variety of flavors: "Open access (OA) refers to unrestricted online access to articles published in scholarly journals, and increasingly also book chapters or monographs. [...] OA can be provided in two ways:

- "Green OA" is provided by authors publishing in any journal and then self-archiving their postprints in their institutional repository or on some other OA website. Green OA journal Publishers[6] endorse immediate OA self-archiving by their authors.

- "Gold OA" is provided by authors publishing in an open access journal that provides immediate OA to all of its articles on the publisher's website. (*Hybrid open access journals* provide Gold OA only for those individual articles for which their authors (or their author's institution or funder) pay an OA publishing fee.)"

The OA initiative was prompted by the excessive subscription prices to academic journals imposed by some big publishing houses, who were taking advantage of a captive market, but can be seen as one aspect of a much wider societal trend toward obtaining free access to information that started with the advent of the internet. To give but a few examples, we again cite Wikipedia and the open data initiatives started in various countries to make all data (geographical, geological, economical, etc.) that have been compiled by public agencies available to the citizens. In astronomy, arXiv and even the NASA/Smithsonian ADS are expressions of this trend, and one can also argue that the developers of freeware and shareware software share the same goal of ultimately giving all available information, and the tools to exploit it, to the end users at no, or very limited, cost. This goal reflects the libertarian spirit of the internet's debut and seeks to provide access to information to anyone regardless of social status, origin, or location.

The main problem that this worthy concept has met is the cost of labor. Free access to information projects do not adequately consider the value of the work that goes into producing reliable services and products, as the following examples show.

---

[6]See http://en.wikipedia.org/wiki/Open_access_(publishing).

- In the case of OA, the cost of quality peer review and publication is minimized by the most outspoken proponents of Gold OA, who advocate community review instead of peer review and self-publishing in open repositories such as arXiv – which also have a cost – instead of journals.

- The geographic resources available freely in open data archives would not exist without the costly public service surveys of land and sea that are necessary to provide the basis for precise and error-free maps and charts.

- Wikipedia, in spite of its obvious success, is hard-pressed to provide the rigorous quality control that is the key to a truly useful encyclopedia.

- Shareware such as OpenOffice represents a working alternative to commercial software suites but does not fully meet the standards set by Microsoft Office.

- ArXiv and the NASA/Smithsonian ADS work well because institutions and agencies provide the necessary funds without asking the final users to contribute to the costs of setting up and maintaining the data repositories.

A conclusion that we can draw is that all of the working "open access to information" projects require an adequate level of funding in order to deliver reliable products, but that these costs are either provided for free (such as the huge code developments by the community behind Wikipedia or OpenOffice), hidden (Open Data initiatives, arXiv, and NASA/Smithsonian ADS), or transferred from companies to the information provider (*e.g.*, page charges for authors in OA journals). But the false notion that these products are "free" still floats in the community and is even relayed at the political level. For example, some national governments make the use of open software mandatory in government agencies but do not contribute to their development.

The European Union (EU) requests scientists to publish their results in OA journals when their research projects are funded by the EU Commission. However, the EU acknowledges the cost of OA and provides funding to the researchers to do so, thus financing publishers via EU-subsidized page charges instead of nationally-subsidized library subscriptions. This can be seen as true progress in some disciplines such as biology where publishers have made considerable profits with hugely exaggerated subscription prices and imposed a very long embargo, but such a top-down move is not needed for the structured and limited market of astronomy publishing. In fact, the existence of EU or national-government subsidized, new OA journals could result in destructuring the current astronomy publishing scene, although this appears unlikely in the near future given the prestige of the major journals and the *de facto* OA to astronomical literature offered by arXiv.

It is telling that after a phase of strong opposition to OA the commercial publishers (in astronomy and elsewhere) are quickly jumping on the OA bandwagon. One reason is quite simply that they expect to make more profit on page charges paid by authors to publish in open access journals than on journal subscriptions, the number of which is constantly decreasing as the public funding of universities decreases. The philosophy behind the push to OA by the EU seems to have generous goals, but can also be seen as a hidden transfer of public funding (university libraries) to private publishing companies (*via* authors), in agreement with the liberal policies of the vast majority of EU countries. Ulterior motives have thus developed in the drive for OA that are hardly related to the libertarian ideals that initiated the OA movement.

Having briefly presented the OA background, we discuss the OA policy of the core astronomy journals. We start with A&A, most sponsors of which are EU members familiar with the EU drive to OA. Currently, the last issues of A&A are in Gold OA for the week following their publication. This allows astronomers who do not have a subscription to A&A to read and download the most recent publications in their areas of scientific interest. Once this short period of Gold OA is over, A&A becomes a hybrid OA journal, since authors can pay a fee to have their article in OA. This so-called OA option is rarely used by authors, which confirms that there is no real need for OA in astronomy. The A&A Letters are in Gold OA at all times, as are the online sections (astronomical instrumentation, online data and catalogs, numerical methods and codes, and atomic and molecular data). The reason behind this policy is to provide Gold OA for those sections that are of interest to a relatively wide part of the community and to promote online publication. Furthermore, A&A is a delayed open access journal in the sense that, after a two-year embargo, all published articles can be read by non-subscribers and are thus again in Gold OA. Finally, A&A is at all times a Green OA journal, since self-archival by authors is encouraged.

The AAS journals ApJ and AJ as well as PASP offer Green OA and delayed OA, with a one-year embargo leading to Gold OA. MNRAS offers Green OA and delayed OA after a three-year embargo. The core astronomy journals provide an OA option at a nominal cost so they are also hybrid OA journals.

It is clear that the pressure exerted by governments and funding agencies on researchers to get them to publish in OA journals will continue growing in years to come, and the traditional journals will need to adapt to this new requirement. The interests of ApJ, AJ, PASP, MNRAS, and A&A are the same in spite of their different business models: we all want the serve our respective communities in the best possible way, and this may provide a valid incentive for a common approach to full OA in the future.

6 Conclusions

We have discussed various aspects of the leading journals in astronomical research to show what publication consists of in our area of scientific work. Although they differ in their details, the operations of astronomy journals are very similar. In particular, the acceptance criteria for published articles are very much the same for the four main journals, a homogeneity that results directly from the small size of our community and the use of the same pool of referees. As a consequence, the impact of the core astronomy journals, which publish about 90% of the global astronomical production, are similar within a factor of less than two (an arguably negligible difference for astronomers). This contrasts strongly with some other scientific fields, such as molecular biology, where impact factors of the various journals range from less than one to more than fifty.

One question that surfaces again and again in our community is whether we still need journals. Why can't we simply put our results on astro-ph, and let the readers decide what is sensible and what is dross? One argument in favor of journals is trivial: who has time to read badly argued papers for extracting the useful information that might lurk in them? Journals act partly as a filter for the publication flow that recycle poorly organized material into better structured, useful pieces of work. The second argument is complementary: as shown above, peer review does improve the papers' science. Authors acknowledge the useful role of referees, although they might be tempted to vilify them at times. Finally, the journals assure the preservation of the validated scientific results that they report, and constitute the scholarly record in a way that other community resources, valuable as they are, do not. As a consequence, the evaluation of scientists by their funding agencies rely in good part on their publication record in peer-reviewed journals.

The validating role of the research results played by academic journals is in fact so important in the scientific process that it should logically lead to recognizing that journals are an integral part of the research infrastructure, on the same level as scientific instruments and archival databases, which would justify financing community-owned journals by the science-funding agencies. Although the profit-oriented approach of some big publishing houses is hardly compatible with the values of the scientific community, academic journals should not dispense with publishers: their role is essential in shaping professional-looking journals that do justice to the exciting research results published in their pages. However, the journals ought to associate with low-profit publishers (*e.g.*, university presses, community-owned publishing houses) and work with them toward establishing fair publication prices allowing for a wide, free dissemination of the scientific results.

THE CORE ASTRONOMY JOURNALS

Claude Bertout[1], Chris Biemesderfer[2], Robert F. Carswell[3], Kim L. Clube[4], Thierry Forveille[5], John S. Gallagher III [6], Paula Szkody[7] and Nathan Vishniac[8]

Abstract. This chapter presents the core research journals in astronomy and astrophysics, as well as the learned societies or consortia that publish them. The authors provide some historical details concerning the origin of societies and journals and explain the journals' areas of interest and editorial policies. The journals' economic models are also briefly mentioned.

1   The American Astronomical Society and its Journals Program

The American Astronomical Society (AAS) was founded in 1899 by a group of leading American astronomers. In the ensuing 110+ years, it has grown to be the largest association of professional astronomers in the world. The AAS is well-respected for its support of astronomical research through significant conferences and publications, as well as for its thoughtful public policy initiatives. The AAS publishes several of the largest and most important research journals in astronomy as part of a publishing program that has been at the forefront of innovation in online communication for decades.

1.1   Condensed History

The American Astronomical Society (AAS) was born at the end of the 19th century. Founded in 1899, its purpose is to help foster astronomy and closely related

[1] Observatoire de Paris, 61 Av. de l'Observatoire, 75014 Paris, France

[2] American Astronomical Society, Washington, DC, USA

[3] Institute of Astronomy, Madingley Road, Cambridge CB3 0HA, UK

[4] Royal Astronomical Society, Burlington House, Piccadilly, London W1J 0BQ, UK

[5] Observatoire de Grenoble, BP. 53, 38041 Grenoble Cedex 9, France

[6] Astronomy Department, University of Wisconsin, 475 N. Charter St., Madison, WI 53706, USA

[7] Department of Astronomy, University of Washington, Seattle, WA 98195, USA

[8] McMaster University, 1280 Main Street West, Hamilton, ON L8S 4L8, Canada

sciences. George Ellery Hale was instrumental in the founding of both Yerkes Observatory at the University of Chicago and the Astronomical and Astrophysical Society of America, which was the Society's original name. The name was changed to the American Astronomical Society in 1914. Hale, who also co-founded the *Astrophysical Journal* in 1895, was skilled at recruiting and influencing talented scientists, and he assembled a group of the most prominent astronomers in the United States when the Society was formed. In the process, he managed to engage Simon Newcomb as the organization's first president (Osterbrock 1999).

In its early years, the Society produced the *Publications of the AAS* to record the transactions of the organization, including abstracts and reports from its scientific meetings (Milkey 2006). Initially distributed in various scientific publications of the day, the reports from the meetings were gathered together in 1910, and the *Publications* appeared regularly until 1943 (Stebbins 1947). In all, ten volumes were published which included abstracts from the first seventy meetings of the AAS. Abstracts from the Society's meetings from 1944 through 1968 were published in the *Astronomical Journal*. Since the 128th meeting of the AAS, abstracts of meeting presentations have been published in the *Bulletin of the AAS*. The *Bulletin* was introduced in 1969 to accommodate the anticipated increase in meetings sponsored by the Society and its Divisions, and a corresponding increase in meeting presentations (Schwarzchild 1969).

The *Astronomical Journal* (AJ) was founded by Benjamin Apthorp Gould in 1849. Gould was brilliant and driven, much as Hale was, and Gould sought to professionalize astronomy in America as had been done in Europe in the preceding century and a half. The creation of a journal "exclusively for the advancement of science" in astronomy was a crucial element in his efforts (Gingerich 1999). Gould edited the journal until 1861, when publication ceased as a consequence of the US Civil War, and again from 1886 until his death in 1896. The AJ was published by the Dudley Observatory in Albany, New York from 1912 until 1941, when its ownership was transferred to the AAS (Hodge 1998).

The *Astrophysical Journal* (ApJ) was founded in 1895 by Hale and James Keeler (Osterbrock 1995). At that time, the journal was owned and published by the University of Chicago, which also owned and operated Yerkes Observatory. Its early editors were always the directors of Yerkes, and its editorial management was the purview of Chicago faculty until 1971. The journal's ownership was transferred to the Society in the early 1970s (Abt 1999).

In the 1990s, the AAS began to utilize several technologies to take advantage of the digital capabilities that were becoming accessible to astronomers. Authors started to submit electronic manuscripts prepared with the AASTeX vocabulary that extends the LaTeX system. Beginning in 1992, the ApJ issued a series of video tapes to present illustrations that were more informative in a multimedia format (Abt 1992). The AAS CD-ROM Series was launched in January 1993 (Abt 1993), and over the five years that followed, nine volumes of digital data were released. In 1998, the journals began publishing the same sorts of multimedia files and data sets online with the corresponding articles (Hodge 1998; Kennicutt 2001), and the videos and the CDs were discontinued.

In 2001, *Astronomy Education Review* (AER) was launched to provide a journal for scholars in the field of astronomy education (Fraknoi & Wolff 2001). Originally published at the US National Optical Astronomy Observatories, the journal was transferred to the AAS in 2009 (Fraknoi & Wolff 2009), where it remains a rich resource for astronomy education research.

1.2 The AAS Today

The AAS is first and foremost a *membership* organization, a collective of members who share common goals and a common mission. Since its founding, the Society's membership has grown to more than 7500 members. Most of the membership is from the United States, although the AAS has a global reputation because of the quality of its meetings and publications. Membership in the AAS is possible in several classes, and is available primarily to research astronomers and other professionals who are actively involved in the advancement of astronomy or a related science. In order to become a member of the AAS, a person must be nominated by a present (Full) member of the Society. International Affiliate Membership is open to astronomers who do not reside in the United States. Applicants either must be nominated by an active Full member or must be members of an affiliated society that requires nomination for membership. See `http://aas.org/membership/classes.php` for more information about AAS membership.

The AAS convenes some of the most productive *meetings and conferences* in astronomy. The Society organizes two general meetings each year – in January and (usually) in June – and there are regular meetings organized by the Society's various Divisions. All these meetings take place in North America, most of them in the US. The Society's meetings are open to anyone interested in astronomical research. The AAS also organizes topical meetings on a range of subjects of interest to the community. See `http://aas.org/meetings` for more information about AAS meetings.

In addition to strengthening interactions through professional meetings, the AAS supports member *divisions* that represent specialized research and astronomical interests. Those specialized areas are planetary sciences, high-energy astrophysics, solar physics, dynamical astronomy, historical astronomy, and laboratory astrophysics. Members of the AAS may join one or more of the Society's Divisions for modest additional dues. See `http://aas.org/divisions` for more information about the AAS' Divisions.

Central to the Society's mission is the publication of scholarly *journals*. One of the goals of the AAS is to publish important, technologically advanced, and high-quality research journals in the field of astronomy and astrophysics at the lowest possible cost to both authors and subscribers. The AAS today publishes five research journals: the *Astronomical Journal*, the *Astrophysical Journal*, the *Astrophysical Journal Letters* (ApJL), the *Astrophysical Journal Supplement* (ApJS), and *Astronomy Education Review*. Reports (abstracts) from meetings of the AAS and its Divisions are published in the *Bulletin of the AAS*.

See `http://aas.org/journals` for more information about AAS journals.

The AAS also has programs to support education, press relations, and public policy. It publishes a number of newsletters on a variety of topics. The Society and its Divisions recognize outstanding contributions to the field of astronomy by awarding a wide array of prizes.

1.3   The AAS Journals Today

The AAS publishes a broad spectrum of research topics in the *Astronomical Journal* (AJ) and the *Astrophysical Journal family* (ApJ, ApJL, and ApJS). Topics pertaining to education research in astronomy and space sciences are published in *Astronomy Education Review*. Each of these journals has an editor-in-chief and a professional staff to support the peer review operation. The journals are published, in print and online, by the Society's publishing partners at the American Institute of Physics and the Institute of Physics in the UK.

While the Society continues to print and distribute journals on paper, the demand for print subscriptions has diminished significantly over the last decade. It is anticipated that the printed edition of the journals will exist only in regional depositories within the next three or four years. The AAS has prepared for the networked digital environment for well over two decades (Biemesderfer 2010), and at this point the Society's attention is focused on the digital forms of the journals. More and more, the print edition serves as a digest or survey of the critical research elements being presented by the authors – elements that are increasingly available only as digital objects.

Under the AAS' stewardship, its journals have advanced in innovative ways over the past half-century, and now rank among the leading astronomical research journals in the world. All the AAS journals accept (and insist on!) original digital source materials from submitting authors; authors' sources are transformed into XML and other scalable standard formats upon acceptance, and the journals' delivery formats, both online and print, are derived from those standard masters. The standard master forms of the articles (not the print, not the PDF, not the web version) constitute the "journal of record", and they are actively managed (curated) for on-going dissemination and preservation.

The AAS journals program partners with external services and initiatives that are crucial to effective management of the program's assets for the benefit of researchers. These partnerships include CrossRef for citation linking and Portico for digital preservation, as well as newer initiatives such as ORCID for contributor identification and management, and DataCite for data set citation. Of course, the AAS journals are engaged with ADS, several of the major astronomical data centers, and the international virtual observatory efforts.

As the journals move ahead, significant investments will be made to facilitate the formal publication of data, following the principles that apply to the traditional scholarly literature. Several varieties of data appear in the journals already, although many improvements and enhancements are planned. In these and other ways, the AAS journals will continue to offer first-rate publications and publishing opportunities for astronomers around the world.

1.4 The Astrophysical Journal

Founded in 1895, The Astrophysical Journal (ApJ) is one of the world's leading research journals. It is owned by the American Astronomical Society and its editorial staff operate report to the society through the Publication Board of the society. The main journal is currently published three times a month, and each issue contains approximately 90 papers. Since 1967 shorter and more urgent communications are contained in the Astrophysical Journal Letters. In addition, there is a monthly issue of The Astrophysical Journal Supplements, which contains longer papers, aimed at a narrower audience, *e.g.* describing methods or instrumental work, or containing large amounts of relatively uninterpreted data. The Supplement Series also is used to publish special issues, containing the results of special projects or space missions. The readers, authors and editors of The Astrophysical Journal come from scientific communities all over the world.

The Astrophysical Journal is a peer-reviewed journal. Submissions to the ApJ are expected to contain "novel and significant" results and are judged according to this standard. Review articles and comments are not considered for publication. The topics covered by The Astrophysical Journal cover the full range of astronomical topics, from solar physics to the early universe. However, we do consider whether or not a paper is likely to reach the most appropriate audience through our journal. For example, papers on the fundamental aspects of General Relativity or alternative theories of gravity are likely to be redirected to physics journals specializing in these topics. Submissions to the main journal are handled by an editorial board which finds reviewers for the articles, mediates in disputes between referees and authors, and decides on the ultimate disposition of papers. It is normal for a paper to undergo at least minor revisions, and not unusual for a paper to be revised two or more times before acceptance. Referee reports are expected within 30 days. In practice this means that it takes about five weeks for a report to be delivered to the authors. The median time between submission and acceptance is a bit more than 80 days. Once accepted a paper is sent to the publisher. The copy editing and typesetting process normally takes 50 to 60 days, at which point the paper is posted to the web. Articles appear on the web while the issue is still being filled, so the numbering of the articles reflects the posting date. Further information can be found at the web submission site (`http://apj.msubmit.net/cgi-bin/main.plex`) including instructions to authors and reviewers, and a link to the ApJ homepage.

The Astrophysical Journal Letters publishes short articles, which are allowed to be somewhat more speculative but should always be of broad interest to the community. The maximum length used to be defined in terms of published page length, but is now specified in digital terms. The refereeing process is somewhat shorter for the Letters; the first report is expected with three weeks and repeated exchanges between the referee and the authors are discouraged. Further information can be found at the web submission site for the Letters (`http://apjl.msubmit/cgi-bin/main.plex`).

The current Editor-in-Chief of the Astrophysical Journal is Ethan T. Vishniac, who has served in this capacity since the Fall of 2006. The Associate Editor-in-Chief is W. Butler Burton, whose term runs concurrently with the EiC's. The Letters Editor is Chris Sneden, who will be replaced in 2013 by Frederick Rasio. Science Editors are appointed for three year terms by the Publications Board of the AAS, and often serve for multiple terms. A list of the current Science Editors of the main journal can be found at http://iopscience.iop.org/0004-637X/page/EditorialBoard.

The Astrophysical Journal is supported by a combination of publication charges and subscription fees. The former are based on an assessment of the digital content of a manuscript, rather than the traditional page charges. Authors are normally expected to pay publication charges, but in special circumstances can apply for a full or partial waiver.

1.5 The Astronomical Journal

Beginning with its first issue in 1849, *The Astronomical Journal* (AJ) has been at the forefront of rapidly and effectively publishing astronomical research. The AJ publishes articles containing original scientific results derived from observations, including descriptions of data capture, surveys, dynamical processes, analysis techniques, and astrophysical interpretation, including comparisons with theoretical models. Research fields extend from studies of the solar system to observational cosmology. For example, recent papers of note in the AJ cover topics such as properties of asteroids, surveys of supernova properties, and characteristics of modern surveys from the ground and space. In addition, the AJ has a historical imperative to include papers on celestial mechanics and stellar dynamics that have the potential for significant interest in the wider astronomical community. As a member of the American Astronomical Society's (AAS) family of research journals, AJ articles are peer-reviewed. Referees are selected and the review process handled by a small editorial team who have a broad perspective on astronomical research. Papers under consideration by AAS journals must present original work of scientific significance and meet AAS standards for scientific ethics. Referee reports for manuscripts that meet these standards generally aim to improve the quality and clarity of the work to optimize communication between authors and the international astronomical community. Most papers undergo at least one round of revision in response to comments from a single referee, and it is not unusual for more extended exchanges to occur between referees and authors. In cases where disagreements arise between the authors and referee, a second referee may be requested. Such requests are generally granted, provided the issues do not involve fundamental or well established astronomical knowledge. The AJ is a publication for astronomers and astrophysicists throughout the world. Results are published in English that has been vetted and corrected by professional copyeditors. Communication across the world is further aided by the journals international subscription base, and the AJ can be accessed at astronomical research institutions throughout much of the world. Our leadership in exploring the capabilities of electronic publishing to more effectively present astronomical information to our readers also improves the communication and archiving of astronomical results. Authors are strongly encouraged to publish the data underlying their research, and we offer online compilations of related articles to further improve the reader experience. To this end the AJ also considers papers that describe instrumentation, techniques or software that are significant factors in understanding scientific results published in the AJ or its sister journals in the ApJ family. Manuscripts are prepared in latex and Microsoft Word formats and are submitted online via our author web site at aj.msubmit.net. Professor John S. Gallagher III, on the astronomy faculty of the University of Wisconsin – Madison, is the current Editor-in-Chief of the AJ. He is assisted by newly appointed Associate Editor-in-Chief Professor Ata Sarajedini from the University of Florida and Associate Scientific Editor Melissa M. McGrath, who is based at NASA Huntsville. Daniel Scheeres at the University of Colorado consults on celestial mechanics papers. Communications with authors and referees on administrative matters are handled by editorial staff in the AJ office located in Madison, Wisconsin.

Like other AAS journals, the AJ depends on a combination of author supplied publication charges and subscription fees. This approach to funding allows us to maintain reasonable subscription rates as well as quick (1 year) open access to our articles. Not surprisingly, many AJ papers have considerable longevity and remain important references for many years.

Please visit our web pages at aj.aas.org for more information, including recent papers, updates on editorial staff members, capabilities, and policies, as well as lists of most read papers.

2 Astronomy & Astrophysics

Astronomers in continental Europe first contemplated the possibility of establishing an international journal devoted to publishing the results of their research in the wake of ESO's creation. The preliminary discussions that eventually led to establishing CERN and ESO started less than ten years after the end of World War II. Once the main protagonists of the war had begun to emerge from the darkest hours, the development of scientific research became one of their top priority. At that time, the resentment of former enemies was still very strong in the general populations of all countries involved in the war, but many intellectuals and politicians were pushing common European projects as a way to counteract the nationalistic mind-set that had led to the two world wars. UNESCO played a decisive role in promoting scientific cooperation among Europeans; CERN was born in 1954, and ESO followed ten years later.

In Europe, astronomy was lagging far behind American research at that time, because the USA had been able to build large telescopes early in the 20th century, while none of the largest European countries, engaged in endless military build-up, as well as in civil destruction and reconstruction cycles, had the means to develop large instruments. As a measure of the abyss between the two continents, one only need recall here that the 100 inch telescope at Mount Wilson saw its first light in

November 1917 while the 193 cm Haute Provence Observatory telescope, which remained Europe's largest optical instrument until well into the 1970s, started operation only in 1958.

The publishing situation in astronomy also reflected Europe's divisions. There were a number of national journals, such as Annales d'Astrophysique in France or Zeitschrift für Astrophysik in Germany, which were devoted to publishing national research in their country's own language. The United Kingdom had a long-established journal, the Monthly Notices of the Royal Astronomical Society, which at that time was serving mainly the astronomers of the Royal Astronomical Society[1]. By the end of the 1950s, frustration was running high among astronomers working in continental Europe because their work had very little impact outside their own countries. In stark contrast, American astronomers had established the ApJ and AJ already by the end of the 19th century as the main vectors of communication for the entire US community. On this front, European astronomy was also lagging more than fifty years behind the US at the end of the second world war.

Stuart Pottasch, who with Jean-Louis Steinberg was a major proponent for merging various national astronomy journals into a single international one, has summarized the early history of A&A (Pottasch 1999, 2011). As he clearly states, the new journal was to be run by astronomers for astronomers, and to this day, forty-three years later, A&A has remained faithful to this objective.

A&A in fact belongs to the astronomers of the countries that sponsor its operations through modest annual contributions of the national agencies that fund astronomy. The sponsoring countries form the A&A consortium. They subsidize and run the Journal's operations, in return for which authors from these countries publish in A&A for free. Each of the sponsoring countries elects one representative to the A&A Board of Directors, the owning and governing body of the Journal. The Directors in turn appoint the A&A Editors, who are responsible for the contents of the Journal and are accountable to the Board.

One should note here that the Board of Directors, although it owns A&A, is not a legal entity. ESO was therefore associated to A&A from the start and provides a legal framework for the Journal by taking care of its financial operation and holding the copyright for A&A articles. Since A&A publishes many papers from ESO researchers, the editorial process must of course remain independent of ESO to allow for an unbiased evaluation of its scientific production, meaning that the connection between ESO and A&A is thus limited to administrative and financial concerns.

A&A is a wide-ranging journal, which publishes new and significant results of astronomical research, regardless of the technique used to obtain them, and accepts papers from any origin; i.e., it is not restricted to serving the sponsoring countries of the Journal. The A&A Main Journal contains research articles and

---

[1]MNRAS has since then become an international journal with two-thirds of its contents originating outside the UK.

research notes, while the Letter section contains short (max. 4 pages) manuscripts of particular timeliness and significance. Review articles and comments are not considered for publication.

Over the years, A&A has grown to become one of the largest astronomy journals in terms of volume (currently more than 20 000 printed pages per year) and is serving researchers from more than sixty countries. Twenty-four of them currently sponsor the Journal; these are not only European but also South American. The current sponsors are Argentina, Austria, Belgium, Brazil, Bulgaria, Chile, Croatia, the Czech Republic, Denmark, Estonia, Finland, France, Germany, Greece, Hungary, Italy, Lithuania, the Netherlands, Poland, Portugal, the Slovak Republic, Spain, Sweden, and Switzerland.

2.1 The A&A Scope

To illustrate the following presentation, we present some graphs based on the 2010 A&A operations. The Journal situation is stable, so that figures given in these graphs are representative of the past few years.

In contrast to other astronomy journals, A&A uses sections to order the published papers. Each section corresponds to a broad scientific area of astronomy or astrophysics to help the reader find more easily the papers that match her scientific interest in the table of contents. As it turned out, the division in sections allows for a natural and progressive transition to online-only publication of the Journal.

The percentage of papers received in 2010 in each of the scientific sections are shown in Figure 1: 19% of the papers we receive concern stellar physics, 16% discuss extragalactic astronomy results, and 13% deal with interstellar medium studies. This distribution has remained fairly constant over the past decade, although we have seen a slow but steady increase in the submissions in cosmology and extragalactic astronomy. More recently, we have been receiving increasing numbers of papers concerning planets and planetary systems, which mainly reflects the current advances in extrasolar planet searches and in planet formation theory. This evolution reflects the driving roles of the ESO VLT in astronomy and of community access to several successful space and ground observatories from X-ray to millimeter wavelengths.

2.2 The Editorial Process

Once a new submission is registered on the A&A Manuscript Management System (https://mms-aanda.obspm.fr/is/aa/) by its authors, its scientific evaluation is handled by an editorial team comprising the newly elected Editor-in-Chief (Thierry Forveille), the Letters Editor-in-Chief (Malcolm Walmsley), the Managing Editor (Claude Bertout, who was Editor-in-Chief from 1999 to April 2012), and seven Scientific Editors (currently Françoise Combes, Andrea Ferrara, Tristan Guillot, Ralf Napiwotzki, Hardi Peter, Steve Shore, and Eline Tolstoy). All editors are chosen by the Board of Directors for their wide and in-depth knowledge of at least one of the various scientific subfields of astronomy and their ability to handle

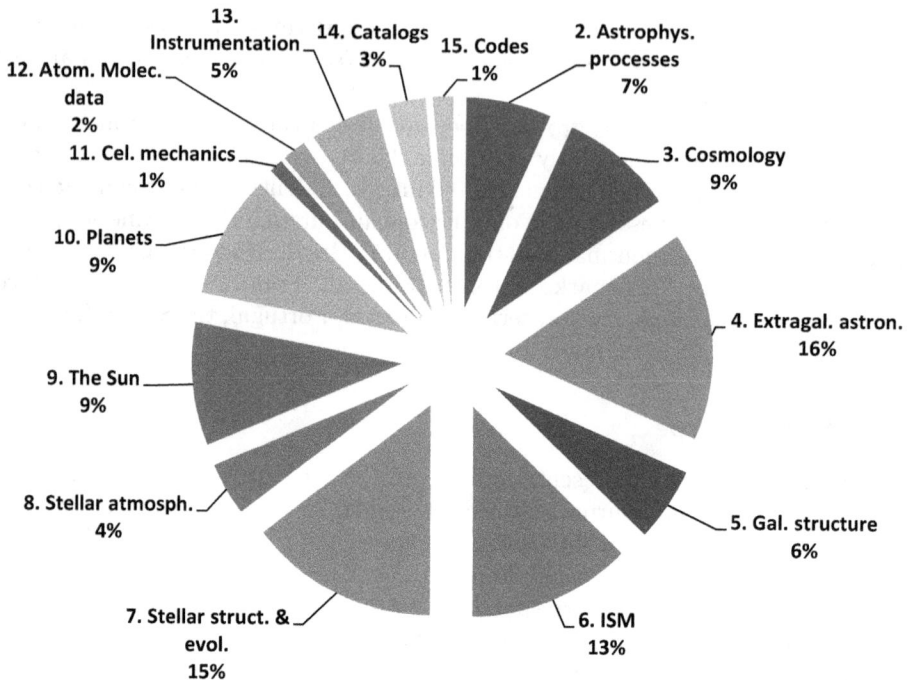

Fig. 1. Percentage of papers received in 2010 in each scientific section.

the peer review with fairness and diligence. The A&A Editors handle both Main Journal articles and Letters to the Editor. They choose referees and overview the peer-review process. The editorial office, located at the Paris Observatory, provides help to the Editors, handles the administrative correspondence with authors and referees, and interfaces with the publishers. The editorial staff currently includes two assistants (Jennifer Martin and Pascale Monier) and three language editors (Lois J. Adams, Claire Halliday and Astrid Peter).

Editors do not restrict the choice of referees to those scientists who work in A&A sponsoring countries. Because astronomical research exists in a framework of global cooperation, the peer-review process must also be global. Figure 2 shows the origin of reviewers who evaluated A&A papers in 2010. Only countries contributing more than 1% of all referees are shown there. We note in particular that 40% of all A&A referees are from North America, where a large fraction of the global astronomical community is working; conversely, the US journals call on many European referees.

Referees focus primarily on the scientific contents of the paper. A&A provides a short questionnaire as a guideline to the various aspects of the evaluation that the editors would like the referee to address. A report prepared by a seasoned referee will not necessarily address all these questions explicitly but does so implicitly

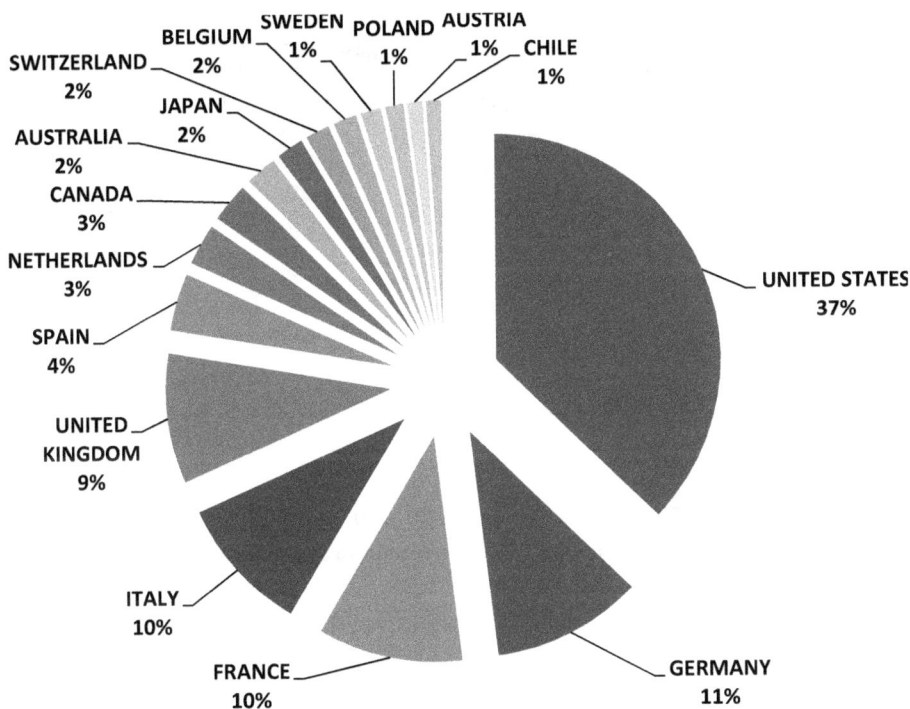

Fig. 2. Origin of A&A referees in 2010.

instead. This is perfectly acceptable, since the questionnaire is mainly intended to help referees with little previous experience of the peer-review process.

The peer-review process, from article submission to acceptance, takes three months on average, but a more meaningful indicator of the time needed for acceptance of a submitted paper is shown in Figure 3, where the fraction of articles accepted in 2010 is displayed as a function of the time between submission and acceptance. About 10% of the papers – presumably the best ones – are accepted within one month, and the median acceptance time is 81 days. Three-thirds of all papers are accepted within four months, but the tail of the distribution extends to one year. These times include the times needed by authors for revising their paper, which is generally longer than the time spent with the referee.

This issue is particularly relevant for the Letters to the Editor, in particular when authors try to argue about the urgency of the publication of a manuscript, even though it took them several months to prepare a revised version following the referee report. In these cases, manuscripts are more likely to be forwarded to the Main Journal, since apparently the authors are not all that convinced of the time factor themselves. Figure 4 displays the cumulative distribution of acceptance times for Letters in 2010 and shows that 36% of the submitted Letters are accepted within one month of receipt, 50% within 33 days, and 90% within two months.

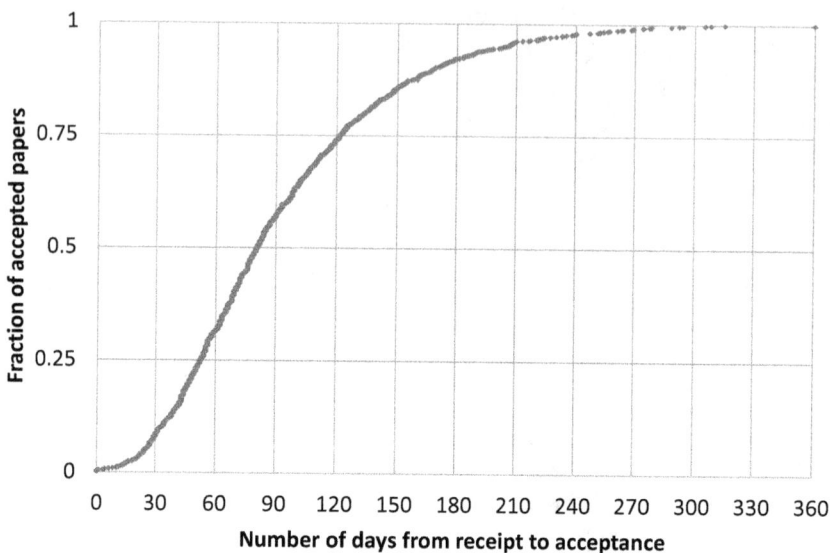

Fig. 3. Cumulative distribution of acceptance times for A&A papers in 2010.

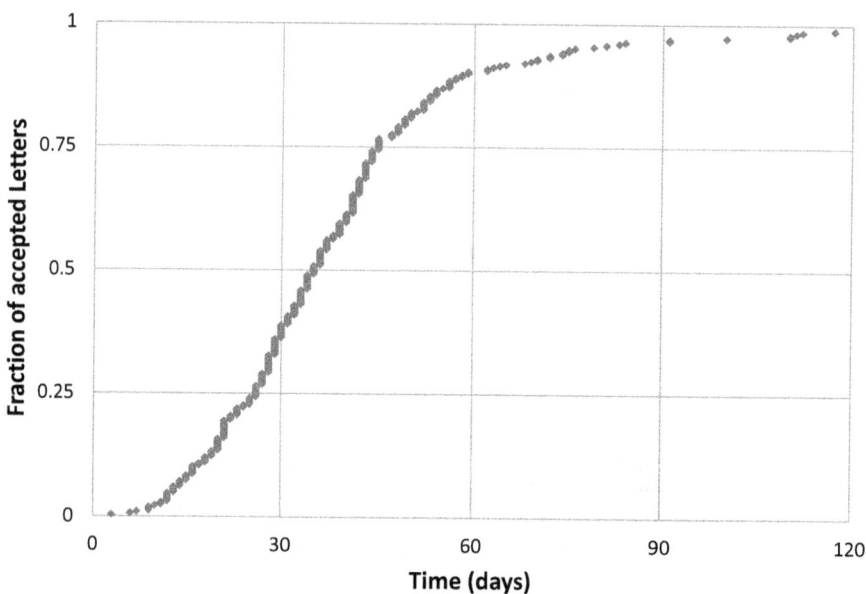

Fig. 4. Cumulative distribution of acceptance times for A&A Letters in 2010.

As expected, this is significantly faster than acceptance times for regular papers. We also note that 29% of submitted Letters were transferred to the Main Journal in the course of peer review, usually because either the editor or referee is not convinced of the urgency of publication.

For more details on the A&A editorial procedures, please see the articles by Claude Bertout and Steve Shore in this volume.

The global acceptance rate of A&A articles is 80%, but there are strong deviations from the mean among the different scientific areas. For example, the rejection rate of articles submitted for the "Astrophysical Processes" section of the Journal is about 40%, because a number of manuscripts received for this section are the so-called "crackpots"[2] that are unsuitable for publication in any scientific journal. The main sections of the Journal (extragalactic astronomy, interstellar and circumstellar studies, structure and evolution of stars, stellar atmospheres, the Sun) have rejection rates ranging from 8 to 17%. Also, papers from sponsoring countries are accepted more often (86% acceptances) than papers from non-sponsoring countries (60% acceptances). This is *not* because editors favor papers submitted by sponsoring authors. Quite the contrary, *all papers go through the same peer-review process regardless of origin.* As a proof of this policy, let us state that one of the countries with the highest acceptance rate is a non-sponsoring country: the UK, with only 7% rejections. Instead, the reasons are simply that (a) some of the non-sponsoring countries that contribute most to A&A submit a number of manuscripts of dubious scientific interest, and (b) some authors from non-sponsoring countries do not have funding for the page charges and are forced to withdraw their manuscripts.

Some parts of the A&A contents (the Letters to the Editor, as well as all articles dealing with astronomical instrumentation, data and catalogs, numerical methods and codes, and atomic and molecular data) are only published online. This reflects the current trend to electronic publication, which allows for savings in the production costs, as well as the changing habits of the A&A readers, who in their vast majority (more than 90%) no longer consult the printed edition of the Journal. We should also mention here that A&A had developed since the early 1990s a partnership with the Strasbourg CDS to archive the observational data published in A&A and make them freely accessible to the community.

2.3 The A&A Economic Model

A&A belongs to the astronomical communities of a consortium of European and South American countries, which sponsor the journal. The peer-review costs are covered through contributions paid by the sponsoring countries to the Board of Directors of A&A, the entity running the journal under the auspices of ESO. Birgitta Nordström is currently the chairperson of the Board of Directors. The production costs of the A&A publisher are covered by the income from subscriptions to the Journal. Nationals from countries sponsoring A&A do not pay for publishing in the Journal, but scientists working in other countries are requested

---

[2] As an example, the commemoration of Albert Einstein's achievements in 2005 brought to us a flurry of papers aimed at disproving general relativity on the grounds of a variety of flawed arguments.

to pay page charges. One priority of A&A is to keep the subscription price (which is set in the contract negotiated with the publisher) at the lowest possible level, while maintaining quality. Since the subscription price is proportional to the number of published pages[3], this implies keeping this number in check. After a strong increase in published pages in the 1990s, A&A went to great lengths to remain at a nearly constant number of printed pages yearly in spite of the increasing flow of information caused by the opening of large observational facilities. This was done by (a) strengthening the editorial policy, (b) publishing some sections only online, (c) increasing the information content of each published page (more condensed page layout, optimized figure size), and (d) publishing some large tabular material only online. As a consequence of these efforts, the total number of printed pages and the subscription price to A&A – after taking inflation into account – has remained almost constant over the last decade, even though the information content published in A&A increased by approximately 30%. In contrast to other astronomy journals, A&A therefore controls its total number of published pages.

Currently, the publisher of A&A is EDP Sciences, a company based in the vicinity of Paris that mainly publishes academic journals and books. EDP Sciences is owned by the French Physical Society, the French Chemical Society, and a French mathematical society. That our publisher is owned by learned societies insures that the company understands the needs and objectives of the scientific community. Besides publication of the electronic and printed editions of A&A, the publisher designs and maintains the A&A website, its accompanying databases, and last but not least, the Manuscript Management System used by all actors in the peer-review system. The publisher also sends the article abstracts to the Smithsonian/NASA Astrophysical Data Service, and allows authors to archive their accepted papers on ArXiV and other reprint repositories.

More details, including guidelines for authors and instructions on how to submit a paper, may be found on the journal website at `http://www.aanda.org`.

3   The Monthly Notices of the Royal Astronomical Society

3.1   The Royal Astronomical Society

In March 1820 the *Astronomical Society of London* (RAS) was set up to promote astronomy. A Royal Charter was signed by William IV on 7 March 1831, and so the name became the *Royal Astronomical Society*. A Supplemental Charter in 1915 opened up the fellowship to women, a few years before women had the right to vote in British parliamentary elections.

Today, the RAS encourages and promotes astronomy and geophysics. This involves publishing journals, organising scientific meetings, maintaining a library and recognising outstanding contributions to Astronomy and Geophysics by the award of Medals and prizes. The RAS also encourages scientists by awarding

---

[3]One should mention here that pages published only online are less expensive than printed pages, which explains in part why A&A is increasingly publishing only online.

grants and post-doctoral fellowships, represents UK astronomy nationally and internationally and makes the case for astronomy and geophysics to government and funding agencies.

The RAS is managed by an elected Council which is headed up by a President and four Vice-Presidents, all of whom are Fellows of the Society. The first president was Sir William Herschel who served from 1821–23. The Societys logo, Herschels 40-ft-long telescope, incorporates the motto of Sir William Herschel: *Quicquid nitet notandum* (whatever shines should be observed). The Society is located at Burlington House in central London and has been there since 1874.

Membership (Fellowship) of the RAS (FRAS) is open to people over the age of 18, and comprises mostly professional astronomers and geophysicists, students and amateur astronomers. Around a third of Fellows are based outside the UK. The RAS holds regular monthly meetings from October to May. In April a week-long National Astronomy Meeting (NAM) is held and is one of the largest professional astronomy conferences in Europe. The RAS publishes two major research journals, Monthly Notices of the Royal Astronomical Society (MNRAS) and Geophysical Journal International (GJI) and also a house journal, Astronomy & Geophysics (A&G) for more general articles and reviews, historical articles, reports of meetings and Society news.

The RAS hosts lunchtime lectures for non-specialists and invites those interested to become friends of the RAS. Friends of the RAS (fRAS) enjoy invitations to social events and meetings in Burlington House, use of the Society's historic library and escorted visits to observatories and other places of interest.

3.2   The Monthly Notices of the Royal Astronomical Society

Publications formed a central activity of the RAS from the very beginning, with the society's *Memoirs* running from 1822 until they were discontinued in 1978. *The Monthly Notices of the Astronomical Society of London* started with a report by the Council to the 7th Annual General Meeting on February 9, 1927. The journal's name changed to *The Monthly Notices of the Royal Astronomical Society* when the name of the society changed to reflect its new status, and it is under that name that reports from the beginning of 1831 were published. This series has evolved with time, and continues to this day as the Royal Astronomical Society's flagship astronomy journal. It is currently published three times a month and, to belie its title further, no longer contains society notices.

While the journal belongs to the *Royal Astronomical Society*, one no longer has to be a member of the society to be able to publish papers in it. Submissions based on new astronomical results are welcome from anywhere.

The scope of the journal is very much as described in Claude Bertout's article describing the publication process in astronomy. In common with other major astronomy journals, MNRAS publishes new results from astronomical research in a wide range of topics. The general requirements are that there be significant new and original material or insight which will be of interest to the astronomical community, and that it be presented concisely but clearly and fully so that the results could be reproduced by others. To ensure these aims are met all published papers will have been thoroughly peer-reviewed.

MNRAS is divided into two sections, the main journal for papers of any length and MNRAS Letters for short communications which merit rapid publication. The short communications may be up to five pages, a limit which is strictly enforced so that both referees and publishers can reasonably be expected to deal with them quickly. Both MNRAS sections are available online, but a print edition is produced for only the main journal. Demand for the printed version is decreasing year by year, but at present (2012) there are no plans to stop providing printed copies for those who want them.

The RAS is interested in serving the community by allowing the results of astronomical research to be widely available as soon as possible. Authors are encouraged to post their papers on the arXiv preprint server whenever they wish to. When this is done is a matter of personal preference - some do so on submission, and some wait until the paper has been accepted and in its final form. Many papers undergo at least minor revision as a result of peer review, so waiting for a paper to be accepted does avoid having to submit corrected versions to arXiv, though of course this does mean the paper is made available somewhat later. In 2011 the median time taken from first submission to acceptance, including refereeing and authors' revision times, was 88 days for the main journal and 51 days for letters.

For the journal papers must be submitted electronically through the ScholarOne Manuscript handling system. For the journal production, accepted papers are typeset, wherever possible, directly from the author's TeX, LaTeX or Microsoft Word file. TeX/LaTeX is the preferred format because of the mathematical nature of the material. Monthly Notices has its own LaTeX class files and TeX macros which simulate the appearance of the journal page and authors are encouraged to use these. Other class files, and other formats, such as Microsoft Word, can also be accepted.

MNRAS is UK-based, but the contributions by authors are welcome from anywhere. In 2011 the journal received about 3000 submissions and accepted 2400 papers originating from 57 different countries. Of these 24% came from the U.K., 36% from other European countries, and 14% from the U.S.A. The leading other countries were Australia, China, Canada, India and Japan, which together made up 26% of the total. A few of the 16 scientific editors are based outside the UK, but most are at UK universities.

Unlike several other journals, there are no page charges, but it is important for papers to be concise: referees or editors may suggest shortening of any that are not, which may lead to delay. The absence of page charges does not mean that MNRAS is vastly more efficient, and nor does it mean that only a minimal service is provided for authors and readers. The basic costs are roughly similar for all major journals, but the financial models are different. MNRAS receives no subsidy from the RAS - in fact the journals produced for the RAS provide part of its total income. Without any subsidy, and with no page charges, the journal costs are borne by subscribing libraries. This does mean their base annual subscriptions are significantly higher than for those for some other journals. For the ∼3800 members of the RAS online access to all the society's journals is included as one of the benefits of membership. This can be particularly advantageous for students, whose membership fees are £ 1 for the first year (see http://www.ras.org.uk/membership).

More details, including guidelines for authors and instructions on how to submit a paper, may be found on the journal website using the Monthly Notices link from http://www.ras.org.uk/publications/journals.

4   Publications of the Astronomical Society of the Pacific

The Astronomical Society of the Pacific (ASP) was started in 1889 when the Director of Lick Observatory (Edward S. Holden) proposed an American astronomical society that would be a counterpart to the Royal Astronomical Society (RAS). He envisioned a society that would continue the cooperation between Lick astronomers and amateurs that began with the January 1, 1889 solar eclipse. Six Lick staff members and 34 others signed the charter membership for the new Society at a meeting of the Pacific Coast Amateur Photographic Association in San Francisco in 1889. Even though headquarters of the Society are still based in San Francisco, the ASP has become the largest general astronomy society, members from over 70 nations. An elected Board of Directors and an appointed Advisory Council manage the Society, whose current goals are to increase the understanding and appreciation of astronomy by engaging scientists, educators, and the public to advance science and science literacy. An Annual Meeting is held each summer at locations throughout the US to provide a forum for the community and to acknowledge outstanding individuals by presenting awards.

The Publications of the Astronomical Society of the Pacific (PASP) is the journal of the ASP and one of the oldest astronomy journals in existence. One of the main goals of the founders of ASP was to disseminate both news and astronomical research among the professional and amateur astronomers. The first circular the group distributed became the first issue of PASP. In the past year, 566 institutions and 416 individuals subscribed to PASP.

The PASP publishes refereed review articles and individual research articles on every aspect of astronomy, as well as short dissertation summaries and conference highlights. The majority of the research it publishes is oriented toward observational work and new instrumentation. Instrumentation articles include new equipment, software for data analysis, and atmospheric monitoring at observatory sites. Because the readership includes astronomy students and technically oriented amateurs, articles on the sociology of astronomy (publishing and job statistics), as well as tutorials and novel approaches to small observatories, may also be accepted. The journal appears monthly with typically 10 to 12 papers and an average size of 1450 pages per year. PASP is generally known for its rapid timescale from submission to publication.

In its 123 years of existence, PASP has had 9 editors, starting with E.S. Holden in 1889. The longest serving was Robert Aitken (1898–1942). The current editor (Paula Szkody) works at the University of Washington. Associate editors Toby

Smith (UW) and Daniel Fabricant (CfA, Harvard) comprise the rest of the editorial group. A Publication Committee that oversees all publications from the society provides advice and support to the editors.

Authors submit manuscripts in either or Word format (with eps figures) through a system called Editorial Manager, which is provided by the PASP publisher, the University of Chicago Press (UCP). Once the paper has successfully passed a technical check, it is forwarded to the Editor, and the peer review process begins within the web interface. After a final revised paper is accepted, it is returned to UCP for the electronic and print publication processes. The median time from submission to acceptance is 2 months and completed papers are posted on a "Ahead of Print" link on the Journal homepage as soon as copyediting is complete (generally a few weeks). Articles are available to the public through ADS after 2 years, but authors are free to post their articles on astro-ph (with posting after acceptance recommended so that the final version is the one posted).

The economic model of the PASP is different from the other US astronomy journals in that any net income from the journal is used by the ASP to promote its mission of advancing science literacy. The review and publication costs are borne almost equally by page charges and institutional plus private subscription charges. Institutional subscriptions are handled through UCP, while individual technical subscriptions are processed by the ASP membership department.

Further information about the PASP can be found at
http://www.press.uchicago.edu/ucp/journals/journal/pasp.html.

4.1 ASP Conference Series

In 1988, the ASP Conference Series was established when Harold McNamara, the PASP editor at the time, recognized the need within the astronomical community for an affordable conference proceedings publication. During the last 24 years, the ASP Conference Series has published more than 450 volumes, 39 IAU Seminar volumes, and five Monographs. These books cover a wide range of topics in astronomy and astrophysics, astronomy education and public outreach, and astronomy software development. The Monograph series includes catalogs of stellar spectra and star-forming regions. ASP Conference Series proceedings volumes help meeting organizers increase the impact of their meetings by sharing the results with a broad international community of astronomy researchers, educators, and students. The Conference Series staff works closely with editors and authors to produce high-quality, low-cost volumes.

The ASP Conference Series publishes about 20 volumes each year and distributes printed and electronic copies of the proceedings to meeting participants, subscribers, and customers around the world. Electronic access to the proceedings volumes is free one year after publication. ASP Conference Series subscribers all over the world receive access immediately after publication. Electronic access to articles is also provided through the popular NASA Astrophysics Data System (ADS), complete with reference and citation tracking, search capabilities, author indexing, and so forth.

The PASP and ASP Conference Series are both among the top 10 cited astronomical journals in the world, and the Conference Series is the only conference proceedings publication in the top 10. In one year the Conference Series website was visited by 150 000 visitors from 195 countries. The Conference Series typically receives requests for more than 9000 abstracts each month, and readers download 1600 articles per month on average. During 2010, 143 000 articles were accessed through the NASA ADS portal. The Conference Series is a popular and relevant source of information for thousands of astronomers and educators around the world. For further details on publishing in the ASP Conference Series, visit the web sites at www.aspbooks.org and www.aspmonographs.org.

References

Abt, H., 1992, The Astrophysical Journal Videotapes, ApJ, 393, 1, doi:10.1086/171480

Abt, H., 1993, Proposed CD-ROM Series, ApJ, 402, 1, doi:10.1086/172107

Abt, H., 1999, The American Astronomical Society and The Astrophysical Journal, in The American Astronomical Society's First Century, ed. D. DeVorkin, American Astronomical Society (Washington DC), p. 176

Biemesderfer, C., 2010, Fully Digital: Policy and Process Implications for the AAS, in Future Professional Communication in Astronomy II, ed. A. Accomazzi, Springer (Heidelberg), p. 83, arxiv:1204.6716

Fraknoi, A., & Wolff, S., 2001, Welcome to Astronomy Education Review, AER, 1, 117, doi:10.3847/AER2001010

Fraknoi, A., & Wolff, S., 2009, Astronomy Education Review Version 2.0: A Welcome and Guide from Your Editors, AER, 8, 010401, doi:10.3847/AER2009017

Gingerich, O., 1999, Benjamin Apthorp Gould and the Founding of the Astronomical Journal, AJ, 117, 1, doi:10.1086/300712

Hodge, P., 1998, Introducing the Electronic AJ, AJ, 115, i, doi:10.1086/300162

Hodge, P., 1999, A Brief History of The Astronomical Journal, in The American Astronomical Society's First Century, ed. D. DeVorkin, American Astronomical Society (Washington DC), p. 165

Kennicutt, R., 2001, Electronic Supplemental Materials for the Astrophysical Journal, ApJ, 557, 1, doi:10.1086/323678

Milkey, R., 2006, The Scholarly Journals of The American Astronomical Society, in Organizations and Strategies in Astronomy, ed. A. Heck, Springer (Heidelberg), p. 241

Osterbrock, D., 1995, Founded in 1895 by George E. Hale and James E. Keeler: The Astrophysical Journal Centennial, ApJ, 438, 1, doi:10.1086/175049

Osterbrock, D., 1999, AAS Meetings Before There Was an AAS: The Pre-History of the Society, in The American Astronomical Society's First Century, ed. D. DeVorkin, American Astronomical Society, Washington DC, p. 3

Pottasch, S.R., 1999, http://www.aanda.org/content/view/103/107/lang,en/; 2011: annotated and reprinted in Scientific Writing for Young Astronomers, ed. C. Sterken (Les Ulis: EDP Sciences), EAS Pub. 49, 23

Schwarzschild, M., 1969, BAAS, 1, 1

Stebbins, J., 1947, The American Astronomical Society, 1897-1947, Popular Astronomy, 55, 404

A Guide to Effective Publishing in Astronomy

HOW TO WRITE A GOOD SCIENTIFIC JOURNAL PAPER

Paula Szkody[1]

Abstract. This articles summarizes the process of writing an article for an astronomy journal that will help lead to straightforward acceptance. It starts with defining good science that merits publication. Then the basic sections that comprise an organized, easy-to-follow paper are described, along with the most effective procedures for making presentable tables and figures.

1 Introduction

When a paper is written well, it moves through the review process quickly and usually requires only one revision. Most journals ask the reviewers to comment on some form of the following questions, with the first one being the most important:

- Does the paper contain significant new results and/or analysis that reflect high scientific standards that warrant publication?

- Is the analysis correct and the paper written with maximum conciseness?

- Is the presentation clear and the English correct?

Thus, the first question to ask when contemplating writing a paper for a journal is "Does my work draw a conclusion that is new, has not been previously published and will advance the field of science in a noticeable way?". Generally, a series of papers based on small increments is not regarded as positively as a long paper that produces a major result from a compilation of the individual studies. For example, a single paper on the lightcurve of a new eclipsing binary would not be well-received, but a paper on the lightcurves of a dozen systems showing anomolous hot spots or wind peculiarities that relate to a different evolutionary path than normal binaries would be significant. Similarly, a study of low-redshift galaxies that adds 20 new objects to a previous study of 500 would not be a significant addition to the literature.

---

[1] Editor, PASP, Department of Astronomy, University of Washington, Seattle, WA 98195, USA

Assuming a positive answer to the above question, and that the analysis uses the latest available software and is accompanied by adequate statistics, the path to a good paper is then straightforward. Whether the interface is Latex or Word, a good paper is generally organized in the manner described below. In all sections of the paper, keep in mind that complete sentences from previous work (even the authors own work) cannot be used in a new paper because it constitutes plagiarism (or self-plagiarism for a repeat of the author's own work) and is not allowed in any journal. Thus, referring to past work requires paraphrasing (summarizing the ideas in different words than what previously appeared in print).

2 General Structure of Journal Papers

2.1 A Good Title

A strong title draws attention to the work described. It should be as detailed as possible within no more than 2 lines of print. For example, in a paper on a study of eclipsing binaries, a title like "A New Study of Eclipsing Binaries" does not give adequate information. A better title would be "UV and Optical Spectra of Eclipsing Binaries in the Young Cluster NGC 1234 Reveals Enhanced Carbon".

2.2 Defining Authorship

In the current era of large surveys, most papers have multiple authors. In large groups, the order of authors can be important, although most journals only list 3–5 before summarizing the remaining authors with *et al.* Generally, any person who has contributed a fair amount of data, or analysis, or who has been a co-author on a proposal for satellite data that appears in the paper should be included in the author list. Anyone providing just one measurement or only a few comments on a completed paper is generally acknowledged at the end of the paper, rather than being a part of the author list. Generally, the first author is the one who put the paper together or did most of the analysis. The rest of the authors can be alphabetical if they did equal work, or listed in order of work contributed. Large-survey papers such as the Sloan Digital Sky Survey used a tier ordering, with those contributing the most in the first tier, followed by an alphabetical listing of those having a lesser contribution or the builders of the project.

Journals provide different templates for author listings. Generally, if there are only a few authors, they can be listed by name with the affiliations directly following. For longer author lists, the affiliation is given as a superscript for which the addresses then appear in footnotes. For ease of contact by readers, it is best to include the emails of the authors in the affiliations.

2.3 A Concise Abstract

As with the title, the abstract is a balance between information and length. The best abstracts are no more than 1 or 2 paragraphs (200–300 words) and highlight the primary reason for the paper, the method of observation or analysis and the conclusions reached[1]. Only firm conclusions should appear in the abstract, with speculation left to the discussion within the paper. There should be no new information in the abstract that is not discussed in the paper. References should not be included in the abstract. It is often best to write the abstract last, after the paper is complete so that all the work can be described in the order it appears in the paper. The abstract is one of the most important pieces of a paper, since it is often the reason for a reader to download the entire paper. The abstract should be understandable to all readers, not just those in the particular field represented by the work in the paper. Keywords usually follow the abstract. They are chosen by authors from a master list that is common to many astronomy journals, and they should describe the main scientific areas covered by the article.

2.4   Main Content

The main body of a paper usually consists of sections labeled Introduction, Observations or Modeling, Results, Discussion, Conclusions and Acknowledgements.

- No matter whether the paper presents observations or theoretical results, an Introduction section is necessary to set the stage. This section requires familiarity with the past literature. It should contain the best, most recent review paper on the subject as well as the latest results that have appeared in the literature. If the work involves one or more particular objects, like stars or planets or nebulae, the discovery data on those objects, plus any relevant literature that impacts the new work should be referenced. All abbreviations used should be defined the first time that they appear. At the end of the literature review, the introduction should clearly point out the purpose of the paper; *i.e.*, why the new study was undertaken and how it will advance the past work.

- For an observational paper, the Observations section should include the detailed information on the telescope, instrument, exposures, observing conditions and the software programs/procedures used to reduce/analyze the data. A summary table here is useful if there is a group of objects observed. For large-survey datasets, the survey can be described in the Introduction but the details of the extracted objects and their analysis would appear in this section. Web sites for handbooks for satellite data can be provided in footnotes. For theoretical papers, this section should describe the details of the software programs and the models used.

- The Results section presents the reduced lightcurves, spectra and model fits to the data. Usually, these results are shown in figures if there is enough new

---

[1] A&A allows for the use of structured abstracts, which help authors organize their abstract with separate paragraphs for the aims, methods and results of the article. Structured abstracts are widely used in some scientific areas, *e.g.*, medical reasearch.

information in each figure. If error bars are not described in the Observations section, they should appear in the figures or the captions of the figures. Fits of data with models are usually also put into tables describing the model parameters and statistical results of the fits.

- Discussion of the data or theoretical results proceeds in the next section. This is where the details of how the data/model improve science should appear and be tied into the past work that has appeared in the literature and referenced in the Introduction. The implications of any problems that emerge from the results, contradictions with other models, etc. should be discussed.

- The Conclusions summarize the findings from the entire study. A clear distinction should be made between the firm conclusions and speculative possibilities. Any further work in the future that is needed to resolve any remaining problems is often included in this section.

- Finally, any people that have contributed some work but that are not included in the author list are acknowledged. Grant agencies that supported the work should also be listed here.

2.5 References

Each journal has a different format for references (some want titles and entire pages while others have shorter lists) but there are some basic rules for all. Most journals refer to references in the text as author and year, not numbers; *e.g.*, Smith (2012). All the references used throughout the paper (including the tables and figure captions) must appear in the reference list, and the reverse also applies in that everything in the reference list must be used in the paper. The list at the end of the article should be alphabetical by last name, and include the year, journal, volume and page. Most journals have abbreviations for journals that are included in their style files (if using LaTeX). If the reference is a book, the title and publisher are given. The number of authors that are listed before a listing becomes *et al.* varies between journals, but is generally 3–5. If there are more than one *et al.* listings with the same first author and year, then they should be labeled as 2012a, 2012b, etc. Papers that are submitted to a journal or in press, or with a preprint available on arXiv:astro-ph can be listed among the reference list, but those in preparation should only appear in the text of the article.

2.6 Tables

As for References, different journals have various formats for Tables, where and how many horizontal lines should appear so the instructions for each journal should be followed. In general, there should be a short (one line title) followed with columns that contain units. Tables should all be referenced in the text with numbers that are in order, *i.e.*, Table 3 must follow Tables 1 and 2 in the text. For long tables, most journals will accept a "stub" table to show the format in the article, with the long table accessed electronically in a format that is machine-readable.

2.7 Making Good Figures

Almost all journals require figures to be submitted in encapsulated postscript (.eps files). This means that they must contain a so-called bounding box that describes the layout. The best figures have large enough points, lines and letters to clearly resolve what is plotted when they are shrunk down to fit within the pages of a journal. Journals generally want CMYK color figures. If figures will be grayscale in the print version and color in the electronic version (the usual option of most authors), care should be taken so that the grayscale shows the intended result. For example, lines can be dashed or dotted instead of using the same line in different colors to differentiate one line from another in a plot. All vertical and horizontal axis should be labeled but labels on top of figures are discouraged. Figure captions should include sufficient detail to understand what is plotted but not duplicate a discussion of the figure that appears in the text of the Discussion section. As with the Tables, all figures must be numbered consecutively and must be used in the text.

3 Cover Letter Information

Along with the manuscript, a cover letter is used to provide a summary of the number of tables and figures and any special instructions, such as which tables are intended for electronic publication, which figures must be in color in both the printed and the electronic editions, and the like. A summary of this type insures that the author and editorial office have the same expectations for how the manuscript will look in the journal.

A Guide to Effective Publishing in Astronomy
© The authors, published by EDP Sciences, 2012

THE EDITOR-REFEREE SYSTEM AND PUBLICATION: AN EDITOR'S VIEW OF THE PROCESS

Steven N. Shore[1]

Abstract. This chapter explains the functioning of scientific journals from the editorial side of the process. Both the history and current functioning of scientific journals are reviewed with a particular emphasis on the evolution of the referee's role. In its current form, the evaluation of a submission is interactive between the three parties – the author(s), editors, and reviewers. The editors serve as the mediators and final evaluators, seeking advice from one or more contacted experts who are in the special position of evaluating the science, presentation, and significance of the work. The chapter explains how this proceeds, and its advantages, pitfalls, and criteria – scientific, archival, and ethical – and how these have evolved historically and consensually. Since referees and editors are also authors, the symbiosis of the process is one of its strengths, since all participants exchange roles.

1 Introduction

The editor plays two primary roles, to choose an appropriate reviewer for a paper and to insure that the process is fair, civil, and in the best interests of both the journal and the parties concerned. Above this there is another function, perhaps less well understood, that of actually accepting the paper for publication. This chapter, written from the point of view of the editors, is meant to serve several purposes: to explain how the editorial process functions in scientific publication, some of its history, and the role of the referees. It is very likely that you will eventually be called on to step to the other side of the publication process, to evaluate your peers' work, so before that happens, it is important to step back and consider why the process exists, how it functions now, and how it changes.

2 Evolution of the Editorial-Referee System

2.1 The Origin of Scientific Journals

The history of scientific communication may seem outside the purview of this chapter but it is relevant to understanding how the different features of scientific

---

[1] Scientific Editor, *Astronomy and Astrophysics*, Dipartimento di Fisica "Enrico Fermi", Università di Pisa, Italy

publication have evolved. And because science as a discipline began in the universities, which were themselves founded in the Middle Ages, I start there. Originally
publication meant depositing the original manuscript of lectures (an *exemplar*)
with a scribal studio licensed by the university (and therefore approved by both
the faculty and hierarchy) who had the right to make, distribute, and, and charge
for, copies on request. These could be any portion of the manuscript or the whole
work, and the material was known to be intellectual product and the rights – that
is, the assignment of authorship – was implicit in the consignment of such material. The authors, who were already being paid for the lectures and examinations,
generally received nothing from the copyists, nor did they have to pay for the
service. It was understood that the rights (such as they were) belonged to the
copyist.

This changed with the introduction of printing in the 15th century. Publishers
went to considerable expense to create their houses, to hire trained staff, and now
also to procure the necessary equipment (*i.e.*, type, the press itself, paper) and
assumed the right to sell the works they produced. But to continue their business,
they were required to be officially sanctioned, and the individual works had to be
submitted to judgment (the *imprimatur* of the local higher clerics) since the work
could now be distributed in a more complete form and was not just for personal
use. A book, once published, had a far wider circulation with many more copies
than a manuscript and each copy was precisely the same. There were attempts at
censorship: the *Index* – an ecclesiastic list of forbidden reading within theological
doctrine – was created in the mid 16th century to inhibit circulation where publication could not be prevented (generally because the publishing houses were outside
of the Papal states, especially in Holland and Venice)[1]. However spectacular the
individual cases of attempted suppression, the literature was unstoppable.

The intellectual and publishing landscape changed in the middle of the 17th
century with the founding of philosophical, that is "scientific", societies. These
received not just sanction but also patronage. They controlled the membership,
being self-governing, which was the forum of communication. The proceedings
of the meetings were recorded, in various forms and to enhance the connectivity
of the different learned societies, these records were exchanged. Thus were born
the *journals*, the very names of which preserve the nature of the enterprise, *e.g.*
Comptes Rendus, *Philosophical Transactions*, various *Acta*, and so on. The society
retained publishers, or assigned the task of publication to individual members (the
secretaries, see *e.g.* Bluhm (1960) for a summary of the Royal Society secretary
Henry Oldenberg), who undertook the production of the reports as the first edited
proceedings. These could be quite elaborate, detailing both the discussions and

---

[1]This did not mean the books were necessarily suppressed, often particularly offensive sections were blacked out as, for instance, in *de Revolutionibus* of Copernicus. Those portions not
contradicting doctrine were permitted reading, even to the extent of allowing the material to
be taught in university lectures. The pursuit of "theological offenders" became more furious in
the late 16th century, Giordano Bruno being the principal example in philosophy, and continued
through Galileo's trial before the Inquisition.

the presentations, and often appeared in annual collections or even extending over several years because of the slow pace of production. Some publishers developed particular specialties, such as an expertise in setting mathematics (with problems similar to Greek or Hebrew texts that required special treatment and types, or music) and often these technical works were published by the same houses as the other humanist productions. The editor overseeing the production often took an active role in production that went beyond verifying the accuracy of the text, one that included communication between the publisher and the authors. This is especially famous in the case of the three editions of Newton's *Principia*, ordered printed by the Royal Society of London and overseen by Edmond Halley. In this role, Halley established himself not only as Newton's drudge but also as an amanuensis, an aid who actually refereed the material, querying Newton at various, often critical, points in the text and requesting extensions or clarifications, or pointing out errors. Robert Cotes served this role in the second edition with even more attention to the details. The editors of the proceedings were, in general, the founders or principals of the societies and as such were in the best position to judge the contents when deciding on publication. But members of the society had the right, by membership, to have their material appear in the proceedings and also in separate works sanctioned by the society. These society proceedings continued through the 18th century to be the main source of current discussion, playing much the same role as the bulletins of meetings – the abstract books – play now. They provided a snapshot of the state of the field at the time. As the membership grew and the finances became more secure, the works appeared with greater regularity and frequency. By the end of the century, there were several dozen in circulation along with an increasing monographic output.

One singularity stands out in the 18th century, the Académie Royale des Sciences (Paris). Unlike the English Royal Society, the academiciens established a separate publications committee, the Comité de Librairie, that served as the adjudicating body for all papers published in the *Histoire* and *Memoires* (McLelland 2003). Even communications to the academy were subjected to critical review before being consigned to the printer and, more important, those outside the academy (called foreigners, not in the sense of nationality but membership) were permitted to submit papers. The significant innovation of the Comité was to create of a set of rules – standards for review and acceptance – that read almost precisely like those later recreated by the modern journals in the 20th century. For instance, a member had to assure the Comité that the paper had not been either submitted or published elsewhere, and papers were refused if they were not sufficiently novel or different from those already published. Expert reviewers were appointed by the Comité to evaluate papers of special interest or in areas in which the members felt less able to judge expertly. This often applied to papers in mathematics and, especially, astronomy. It was not enough that a result be novel, however, it had to be presented in a consistent, complete way that permitted its evaluation. Priority disputes, plagiarism, and proper citation practice were all considered in the criteria, any of which could suffice to reject a paper even if it had been read before to the Académie. This, unlike the Royal Society, presentation at a meeting did not insure the eventual appearance of the complete work but it was a minimum criterion (that is, public presentation).

2.2   Enter the Referees

This changed in the 19th century in several ways, which are fundamental to our world. First, new institutions and societies were founded, more specialized in their nature and more democratic in their membership that did not require particular credentials other than the scientific activity itself and were not class-based, as was the Royal Society. The Manchester Lunar Society, for instance (so named because it held monthly meetings) included practical engineers (*e.g.* Watt, Wedgewood), and the Royal Institution (founded mainly for public outreach, as we'd now put it) opened scientific discussion to the literate masses who began to take a more serious interest in science. This new audience also desired outlets for their interests, and besides more general forums such as the Proceedings wanted their own, less "controlled" (for that read class-laden) outlet. This market was supplied, very soon, with private journals, such as A.L. Crelle's *Journal für die reine und angewandte Mathematik* (known almost universally as *Crelle's Journal*) in Germany and the *Philosophical Magazine* in England. These were under the control of the editors themselves, not any society, and since their funds were supplied (for profit) by subscription they needed to have the highest quality, most interesting material possible[2]. Even among the oldest, best-established publications, there was a growing interest in reaching new readers, by extending the base.

The editors of the *Phil. Trans.* and *Phil. Mag.* were among the best in their field but, given the diverse nature of the material and the ceding of content control to the editors by the various organizations, they found it necessary to get additional advice on the submissions. This was the beginning of the formal refereeing system. The Royal Society, that had maintained a similar body to that of the Académie, the "Committee of Papers", changed to independent referees in 1831 whose function was to independently review the suitability of papers for publication and formally submit a report to the Committee (Anderson 1993). For the *Phil. Trans.*, for example, there are extensive records of the referee reports. For instance, Jackson & Launder (2007) discuss the reports on various papers by Osborne Reynolds (he of the Reynolds number) for which the editor, G.G. Stokes (he of the theorem), maintained a close circle of advisors who could judge papers and, in anonymous reports, recommend acceptance or rejection of the submission including Maxwell, Rayleigh, and Kelvin. The final decision, ultimately, rested with the editor but the reports were the fundamental element in the decision. While this also included the judgment of the editor himself (alas, the gender was

---

[2]There are several of the original 19th century physics journals that, having passed through a number of transformations, are still published, *e.g.* the Italian journal *Nuovo Cimento*, founded and originally edited by C. Matteucci and F. Piria, and the German *Annalen der Physik*, founded in 1790 and most famously edited by J.C. Poggendorff and P. Drude, covered the physical and mathematical sciences (including chemistry).

unique at the time), it was more usually one of the reviewers. The journals also served as proxy meeting sites, publishing commentaries, rebuttals, exchanges in public that would normally have been accessible only in meetings[3].

The next innovation was the weekly and monthly scientific *magazines*, again purely commercial ventures undertaken with adventurous publishers. There were many, most of which failed rather quickly because of the high production costs relative to the subscription incomes. A few are still with us. *Scientific American*, in various incarnations, concentrated on practical matters and brief squibs of information and curiosities. The magazine we now see is much different from that of the 19th century, which more closely resembled *Popular Mechanics* or *Popular Science* (the *American Scientific Monthly* was more serious). The only one to survive the shakeout of the end of the 19th century to remain in publication, is *Nature*, founded by N. Lockyer in the 1860s with the express intention of presenting novel results, as brief communications, along with news and commentary (Meadows 2008). Again, it was the editors who made the decision on what to publish. The American Association for the Advancement of Science (AAAS, the national counterpart to the British Association) began publishing *Science*, the American survivor of this shake-out period, but it was a membership journal so not subject to the same fortunes as the other commercial ventures.

Another feature of the end of the 19th century was the rise of professional organizations, for instance the American Physical Society (APS), the American Astronomical Society (AAS), and the American Mathematical Society, whose explicit aim was to organize and represent a group of newly minted professional scientists. These were not merely groups of amateurs or enthusiasts, they were, in a real sense, the pioneers of the profession. Recall that before the middle of the century even the professionally employed "scientist" did not exist, much less "physicist"[4]. Physics departments, indeed departments and faculties of science, were not founded until the end of the century (although observatories and laboratories did exist separately, as did museums and collections in natural history and botanical gardens that served medical faculties) and the research and educational activity of academic scientists was organized around single professors and institutes. In such an environment, which persisted in Europe far longer than in North America, the various national scientific associations such as the British Association for the Advancement of Science, and the AAAS, mirrored this university-based

---

[3]To give you an idea of who these editors were, and why the reputations of the journals grew rapidly, it helps to look at a very partial list of names: Volterra, Mittag-Leffler, Liouville, Rayleigh, J.J. Thomson, Tyndall, Drude, Wien, and Planck. All had made – or were at the time of their editorship making – outstanding scientific contributions while also guiding by their selections the direction of entire areas of research. A unique view into this last feature is provided by recent studies of the Planck-Wien editorial correspondence for the *Annalen der Physik* (Hoffman 2005; Pyenson 2005).

[4]It is enough to note that the reigning classic English language monograph on mechanics in the second half of the 19th century was entitled *Treatise on Natural Philosophy* by Thomson (Kelvin) and Tait and retained that title to the end of that century. The Cambridge Philosophical Society centered on mathematical physics and mathematics, yet did not explicitly identify either in its name or journal title.

hierarchy, being organized in sections. Thus, the reports of such meetings contained the physical, biological, and mathematical sciences but did not appear as separate fascicles reporting the workings of specific sections. The societies were usually founded, like the Royal Society, along with dedicated journals (for instance, the *Physical Review*, *cf.* Hartman (1994) for the history of the APS journals).

2.3   The Vintage Astronomical Journals

In astronomy, the observatory publications were internally reviewed before material was published, and these appeared in series exchanged among the observatories internationally. The *Monthly Notices of the Royal Astronomical Society (MNRAS)*, similar to the Royal Society *Proceedings*, presented the communications from the meetings but extended these to submitted articles when submitted through a member (who essentially vouched for its contents). The publication, *The Observatory*, was separate, an originally less technical, non-society publication that nonetheless reported on the proceedings of RAS meetings and often included transcripts of the lectures and debates. Perhaps the most famous example of which is the exchange between Chandrasekhar and Eddington on relativistic degenerate gases, see Wali (1991) and the original report, available on the *ADS*. Another example was F.W. Argelander's *Astronomiche Nachrichten* (hereafter AN), which, while publishing observational particulars (for instance, the announcement of Galle's discovery observations of Neptune), also included theoretical material and short commentaries, but avoided news items or meeting notices.

The landscape changed again with the founding of the *Astrophysical Journal* by G.E. Hale in 1895 and the *Astronomical Journal* by B. Gould and B. Pierce a few years earlier. The *AJ* was an American answer to *AN*, and less so to *MNRAS*, being a venue mainly for observational results, especially in astrometry and celestial mechanics, and reflecting the intended audience. It was important that neither of these new journals required communication by a member of responsible editor, both established an open submission policy, which required some form of refereeing in the absence of "pre-endorsement". It was paralleled by a journal with a familiar name, but *very* different character, *Astronomy and Astrophysics*, a short-lived journal that explicitly included the new discipline of *astrophysics* in the title as a polemical provocation. It was Hale's intention to promote this new science (see Wright (1966); the details are found in the centennial volume of the AAS and also in the online history at http://www.aas.org/had/aashistory/). The journal was not explicitly linked to the society, in part because of the conflict over the naming of the new professional organization – the exclusion of the word "astrophysics" from the organizational designation, see Berendzen (1974) for a particularly detailed study. The observatories continued their own series for a very long time even with these alternate venues since they did not have the limitations of profit and space[5].

---

[5]The *ApJ* and *AJ* were founded *before* the community joined in the society-forming mania that dominated the last third of the 19th century, when professional societies were formed to promote the sciences and open the membership to those outside the ranks of the academies.

In founding the *ApJ*, Hale made two interesting decisions. One was to assemble an international advisory board, a very striking attempt to legitimize the journal from the start, including European scientists. The second was to emphasize a technique, spectroscopy, that he felt was the defining distinction between the more traditional areas of astronomy and the possible new science, the physical content of the study of celestial objects. A few spectroscopic societies existed or were sections of existing societies, such as the Italian Society of Spectroscopists, but the *ApJ* was an independent body. Since at the time Hale was at the University of Chicago, itself recently founded, he successfully negotiated the founding with the university press and the ApJ became Hale's personal outlet and the university's property. It was not long afterward, however, that Hale moved to California to realize the construction of Mount Wilson Observatory (later, of course, Palomar) but the journal production remained at Chicago with two offices. The editors were Hale at Mount Wilson and Frost, who succeeded him as director of Yerkes. This highlights the unusual nature of the *ApJ*. The editors were at observatories and, as such, controlled their staff's output but also made it simple for papers to be submitted by walking down the hall and handing it to the principal. Refereeing was introduced but not always used, observatory staff had a priority in the publication, accounting, for the predominance of American authors. Multiple publications were tolerated, some of the papers appeared in abbreviated form in, for instance, the proceedings of the National Academy of Science (prominent examples being the publications by Hubble, van Maanen, Shapley, Hale – who was also one of the NAS founders, surprise!), but these were rapid communications (limited to only a few pages). There were a few conference proceedings, but rather few (the Michelson-Morley experiment conference was one of the few of this sort). Translations of foreign papers were accepted[6]. Laboratory studies were given a pride of place, and series were encouraged (both for the science but also because it was a steady supply of papers on related subjects, some of which were later collected into stand-alone volumes, the most famous instances are Chandrasekhar's monographs[7]). This editorial structure underwent a radical change in the early 1950s. Hale was gone, Frost was retired, W. Adams was at the end of his career, O. Struve had completed his term (also then at Yerkes), and the journal's financial status was perilous (It is the case even though you as authors do not think much of this, that the survival of a journal – even the *arxiv* – requires support.) This brought some of the most dramatic changes in the journal structure and, by extension, established the modern form of journals in our field. S. Chandrasekhar was persuaded to take over the sole editorship when W.W. Morgan resigned and the journal was transferred to the AAS – under the encouraging presidency of M. Schwarzschild

---

[6]This extended a practice from the 19th century of reprint translations, *e.g.* Taylor's *Scientific Memoirs*, and the practice often followed of multiple publications in different languages, *e.g.* Zeeman, Fermi, and others, extending through the 1920s.

[7]An entire issue of the *ApJ* contains what eventually became Chandra's book on stellar dynamics, with a few reprinted papers with, *e.g.* Von Neumann, added. This is an unusual example but it did happen. In some journals now, for instance *ApSS* and *Nuclear Physics*, the volumes are conference proceedings and that also appear as separate books.

– which retained copyright, being held for the authors. To ease the financial situation, nominal page charges were introduced (this at a time when the *Physical Review* was paying referees) paid by the authors or by their home institutions as a supplement to the subscriptions. The AAS also provided funding to maintain the editorial offices, and some support also came from the Chicago press.

My reason for this long background discussion is to explain how the modern journals came to be structured as they are. Claude Bertout provides the details for *Astronomy and Astrophysics*, whose history is even more complex, but it is at this stage that you will see what an editor actually does. Chandra was motivated, as Hale had been, by a particular intellectual program, to increase the presence of theory in the journal. As such, by his own output he encouraged a flow of new papers in radiative transfer, hydrodynamics, stellar dynamics and galactic structure, interstellar medium, cosmology, and new areas of astrophysics such as radio astronomy, while maintaining the well-established areas of solar and stellar physics, laboratory studies, and fundamental properties such as photometry. Wali (1991) reports that he personally refereed about 1/10 of all submissions and had his own papers reviewed in the observatory. *Letters to the editor* were so designated because these were reviewed personally by Chandra without generally being refereed, and comments (short communications) were also included. By the mid-1960s the *Supplements* were added to provide a separate outlet for longer compilations or semi-monographic contributions that were soon to largely replace the observatory publications (but these too had evolved into journals, *e.g. Bulletin of the Astronomical Institutes of the Netherlands (BAN), Bulletin of the Astronomical Institutes of Czechoslovakia (BAC), Astrofisca Norvegica,* and *Acta Astronomica.* Only a few, *Astronomie et Astrophysique), Zeitschrift für Astrophys.,* and still *AJ* followed the same path as the *ApJ*, often with the same centralized editorial structure and refereeing methods. A few new, specialized journals were founded in the 1960s, especially *Solar Physics* and *Icarus,* whose copyright was (and still is) held by commercial houses. As far as I know, among the journals started in this period, only *Astrophysics and Space Science* (the particular project of Z. Kopal, founded as a deliberate competitor to A&A) sought to retain the broad-scope approach of the *ApJ* but with a far looser refereeing structure. The loose board of advisors was replaced by the AAS with an elected publications board that, for the first time, included professional librarians along with working scientists who could advise on bibliographic and archival matters. This last step was vital, for example, to evaluations of new technologies. The board, for the AAS and for *A&A*, operated (and operates) autonomously but doesn't enter into direct editorial matters. These are left to the appointed scientific editors with the result that the functions essentially duplicate the structure envisioned in the 18th century by the *Académie Royale*[8].

---

[8]The International Astronomical Union (IAU) has played an important role in the standardization of the literature since its founding in the early 20th century. While it publishes no journals itself, its presence has been felt throughout the community through its roles in nomenclature and international debate. For instance, specific classification systems, photometric and spectroscopic

With the American post-WW II technological domination of astrophysics, especially the large observatories (not only Palomar but also Kitt Peak, CTIO, NRAO, and the space program) and large computing capabilities, *ApJ* became the focal point of communication in the field by the end of the 1960s. On retiring at the beginning of the 1970s, Chandra was succeeded by Helmut Abt[9], who took several editorial steps that again changed the face of the journal. The *Letters* was separated from the main journal, explicitly providing more rapid communication in an abbreviated format and under the separate editorship of A. Dalgarno at Harvard[10]. The supplements changed from irregular to regular publication, still under the main journal editor, and the format changed to increase the size of the page and the number of pages that could be published, and gradually increased the frequency of publication (which had been steadily evolving in the preceding years). The refereeing base was extended, including younger scientists, and the journal experimented with alternate forms of publication (videos, included as VHS cassettes, CDs, and ultimately online publication). By the time Abt ended his editorship, in 1998, online publication was the rule, electronic submission had been standardized along with a database for editorial control, and – and this is perhaps the most important innovation by the publication board, a group of scientific editors had been appointed.

3 The Scientific Editorial Process Now

This now brings us to the present and the point at which I can begin to explain the work of an editor for *A&A*. The history of the journal and its organization and governance are covered by Bertout so I will leave those aside. I only note that a main difference with society publications is the self-propagating nature of the A&A publications board, whose members are selected by the individual national communities and not by election from a membership, a very significant difference. In general, the scientific editors are chosen by invitation or application from the community. For the most prominent journals in our field, those with the longest histories and visibilities, the choice is by an autonomous representative body[11].

---

standards, even the definition of planets – perhaps the most controversial public decision ever taken by the IAU – are reflected throughout the literature.

[9]One of Abt's major contributions, before joining the journal, was to produce – on his own initiative – the first analytic (keyword) index for the *ApJ*. The journal principal copyeditor at the time, Jeanne Hopkins, also compiled the *Glossary of Astronomy and Astrophysics* based on her notes from discussions with Chandra, who wrote the forward.

[10]The rapid publication format of *Letters* had been pioneered by the *Physical Review* during the early 1960s. The explosive output of accelerators, developments in particle and nuclear physics, and the need to rapidly disseminate the results to the wider community produced a separate editorial structure and policy. Some papers were even put on "fast track", being published without review (but explicitly noted in the published version) to speed up the process.

[11]Some journals are, however, almost "personal projects" of the editors – who are often the founders – but in our field there are few of these. There are also some journals founded by the initiative of commercial publishers who decide themselves who to invite. In some cases the editors are nearly full time, salaried employees. In other cases there may be a unique editor.

The scientific editors (hereafter SEs, also called "associate editors") are distinguished by their research specialties and assigned papers by the editor in chief (for the main journal and the Letters) in those – and related – areas for oversight during the review process. The number of editors varies enormously among the astronomical journals (*e.g.* there are 16 scientific editors for MN, 21 for ApJ, and 10 for A&A) and even more among those in the physical sciences[12]. When a paper is registered, and the necessary administrative details are attended to, the paper is sent to one of the SEs as a new submission. The paper has already passed a first gate, it is been deemed sane, that is "not crank". You would surely be amused to see some of these but they are not a major concern. The obvious bizarre papers are returned to the authors as inappropriate for the journal with a civilized cover letter (not that this excludes the possibility of threats and appeals by the offended authors). But even a paper that appears to be of marginal relevance to the scope of the journal are now sent to the SEs for comment before a decision is made on whether to send it out for review.

3.1 The Editor

The first SE review is a step that needs some explanation. It is not necessarily a judgment by the SE of the correctness of the result, although that comes into play occasionally. Rather, in a world filled with journals that may be more specialized (prestige aside, that is not the point of this decision), papers may be returned because they are better suited to some other venue. This can reflect the evolution of the field – there are specialized journals in almost any area, from general relativity to celestial mechanics to radiative transfer – and given the competition for journal space it is better to send back a paper that will find an audience elsewhere if it passes muster, than overly limit the access of the journal to more relevant papers. More on this point later. There is another, even more important consideration: every paper that goes out for review takes time, freely given as a service to the community, from a working scientist and this is a precious resource that should not, and cannot, be abused. There may seem to be too many astronomers in the world, when bureaucrats tally up their budgets, but the truth is that there are not that many and there are a number of journals, and journal editors, vying for their attention. Every paper under review for the journals in our field is read by the editor and by the referee, often several times along with the responses, and this requires a huge dedication of time and attention by all concerned. An author

---

[12]It is important to note that the astrophysical journals are not subdivided into parts such as, *e.g. Physica*, *Physical Review*, or *Journal of Geophysical Research*. Instead, although there are more specialized journals that address readership in subfields, such as *Journal of Quantitative Spectroscopy and Radiative Transfer*, *Icarus*, *Solar Physics*, and *Celestial Mechanics and Dynamical Astronomy*, the principal journals have chosen to present a more unified perspective and accept papers in a very broad swathe of the research subdisciplines. In this, we are unusual. I know only one field, *Journal of Fluid Mechanics*, that maintains the same very wide scope by deliberate design, similar to Hale's, by the founding editor, G. Batchelor. Batchelor also left a remarkable legacy, his editorial reminscences (Batchelor 1981).

may think it is something just dashed off, if you have never written a review, but a single paper may require days, or weeks, of time by those who will never receive more than an acknowledgment (at most) or remain anonymous to the community, known only to the editors (at least). That the reviews are anonymous, except when explicitly waived (see below), may seem like a way of hiding behind a comfortable mask while delivering a killing stroke to a paper but the editor is there to insure that this does not happen and this is the greatest benefit to the community of the SE structure.

3.2 The Referee

The referee has, then, several responsibilities. One is to the community. As a privileged reader of a submission, both as evaluator and as a fellow scientist, the referee sees the paper – unless it is been placed on the *arxiv* at or even before submission – before anyone else except the editors and is thus party to information not otherwise available. This can put a reviewer in a delicate position of conflict of interest, if the research is too close to her/his own work, but it is almost always (and those few exceptions will be discussed further in a moment) signaled at the time of the request[13]. The process begins with the editor sending a copy of the abstract to the prospective reviewer, and if properly written, it will contain more than enough information to permit a potential reader to decide whether to accept the task. But this is also privileged information. We know well that a colleague, or competitor, can extract a wealth of information about a result from just a few key words. An abstract that is properly written for later introduction of the paper to the readership will also convey enough information to an unethical reader to jump the process and try to publish first. One of the most impressive characteristics of the astronomical community that this rarely happens. It is more likely that someone will see a talk at a meeting[14].

How the review process starts and is monitored and mediated illustrates the centrality of the editorial board in the process and highlights the importance of *scientific* (as opposed to administrative) editorial control of the refereeing process, The editor in chief receives the paper for consideration when all bureaucratic requirements have been satisfied and, on reading the paper, selects the appropriate SE for overseeing the evaluation. Some journals have a schedule for distributing

---

[13]Large collaborations, *e.g.* Fermi, LIGO, H.E.S.S., and SDSS, frequently have an internal refereeing process that vets papers before submission. This substantially improves the quality of the version sent to the journal but cannot replace the "fresh" review provided by an external referee. Often something will be spotted by a disinterested reviewer that had been missed by the affiliated readers. Remember, within large groups there is always the danger of a sort of "private language" or "group-think" developing, a common way of approaching the material that is hard to step out of.

[14]An anecdote is in order here to emphasize the difference between how referees *do not* behave. Recently, "preliminary" data presented in a slide at a meeting has been photographed by a listener using a cellphone and then sent directly to competing groups, with a paper being submitted before the end of the meeting discussing data that has yet to be submitted. One of the editorial roles is also to have a sense of when this is happening, but more on that later.

the papers to the editors, for others it is a continuous process. The arrival rate for papers is certainly *not* a random variable, since there are strong correlations with the outside world of deadlines (for, *e.g.* observing time proposals, grant applications, fellowships, theses, vacation periods) and these affect all journals. There is also a strong "Monday Morning effect", when papers that have arrived after close of business on Friday may have to wait for assignment depending on how the editorial staff is scheduled. Again, this differs between journals but also affects later contacts with possible reviewers. So assuming all goes smoothly, a paper may arrive in the mailbox of an SE up to a few days after the authors have been told that the process has been started.

The criteria for selecting a referee are flexible, and this is where the process is most vulnerable – at least in principle – to the whims of an editor. The wrong choice of a reviewer colors the whole process, but it is the task of the editor having sufficient background in the field to decide who can both evaluate a paper on its scientific merits and provide an external reading. And for this reason the editor is always visible in the process, the one who ultimately takes responsibility. The choice is made on several different levels. Along with the paper, the authors can state a preference for which SE they think would be the appropriate choice for overseeing the evaluation. They can also suggest potential referees, or signal those who they prefer to see excluded.

This information is considered at all stages but there should be explicit reasons for avoiding an individual or group and/or additional information for those in the preferred category[15]. Close colleagues are excluded, or at least close collaborators, usually based on publication history. Competitors are also excluded, although more on this point in a moment. The narrowness of some subfields can make this step particularly difficult, especially for large collaborative papers that can virtually preempt a whole field. But at this point, the first editorial reading of the contents of the paper is important because there may be some outside the precise area of study who are perfectly capable of evaluating the paper without being parties to the various controversies. The bibliography is important in this regard, it may indicate some of the connections that the editor does not know, or indicate who might be outside the group but not compromised by rivalry. The *ADS* is another important source. Tracing back a citation history for a possible referee can reveal connections, possible biases, or other problems. It can also provide useful additional context to the editor[16]. Depending on the number of papers being dealt with, an individual editor may take some days to begin the contacts (an invisible step, the authors see only the date of registration of the paper).

---

[15]It happens, occasionally, that authors suggest collaborators as reviewers, for instance. This is *not* a very good idea: not only does it look ridiculous, it also indicates an irresponsible approach to the publication process. In other words, be thoughtful when making any reviewer suggestions.

[16]This is neither an instantaneous process, nor can it be automated (well, it can be but I am reminded of the play on words of a sign in a laundry "we do not tear your clothes by automated machinery, we do it carefully by hand").

There are also more than one editor for the journal and even SEs can collide, when the ideal reviewer for one paper may already have recently accepted the request of another editor to take a paper, or may have been asked recently enough that a new request would be an abuse. Some colleagues, dissatisfied with some previous experience with an editor or a journal may systematically remove themselves – implicitly or explicitly – from the process, further reducing the available pool (although this does not happen often it has and does occur)[17]. At times, even if unable to take a paper, a contacted reader will suggest an alternate referee. Like the author's suggestions, this is taken under advisement but not always followed. It is, however, an excellent way to get younger scientists into the refereeing pool.

An important distinction in both the selection and review process between the Letters and the main journal is the criterion for acceptance. This is similar for most of the astrophysical journals. A Letter must be both timely (read "urgent") and particularly interesting. The latter is, of course, a value judgment but it can be substantiated and, indeed, must be in any evaluation. For some journals, the paper may be more speculative, perhaps even on the borderline of being "far out", but it must not contain any obvious errors. The timescale for evaluation, usually about half or less than the nominal time allowed for a main journal submission, means that the process is somewhat looser but no less serious. A Research Note requires less contextualizing than a regular paper but does not have the urgency of a Letter. Actually, in this age of instantaneous communication of preprints, the Letters have an odor of anachronism. They continue to be regarded as, somehow, more prestigious than the main journal, a hang-over from earlier times[18].

When the person has been selected, the contact process begins. This can be long and, from the outside, may seem very inefficient. To an author consulting the submission system (in our case MMS), to see a list of contacts of contacts (by number only, i.e. referee 1, referee 2, etc.) doesn't necessarily inspire confidence. It may appear that the editor is "scraping the bottom" to find a reviewer. But we are all busy and to review a paper takes time. There are always those who do not respond the first time, or the second (and editors reserve a Dantean ending for them), but other factors can enter. Spam filters can block messages without

---

[17]Also, keep in mind the number of submissions per year to the principal journals, including those in other areas of physics, informatics, and even engineering, for which any scientist may be asked to review. There are literally tens of thousands of such requests in circulation each year, an almost terrifying volume of papers! It should, therefore, not be a surprise that at times a paper can take a number of requests, to very qualified scientists, to finally end up with an appropriate reader who *is* available.

[18]This is particularly true for the physics community that regards the *Physical Review Letters* as the place of choice for "important" results even though any paper submitted there can also be placed online and the statistics of acceptance are quite skewed by the evaluation process. Let me explain. A paper can easily be rejected after a single pass if there are two (or more) reviewers, as happens with some journals. The revised paper, resubmitted, may make it through easily yet the statistics for acceptance – the impact factor – will be distorted by the process. Main journal submissions usually go through at least one, sometimes two rounds. An unscrupulous journal could easily increase its statistical standing by changing this procedure. Fortunately, most journals are not yet succumbing to this temptation.

sending notification to the journal[19]. Assuming the potential reviewer does reply, the response may be that the request comes at a bad time, or that the paper is too far from her/his current interests, or even that there is a possible conflict of interest in the review. The she/he may suggest someone else as a better suited for the paper. Remember, there are a number of journals and as a round number we have about 10^4 potential reviewers, around 3000 papers per journal per year, and about a dozen journals just in astronomy (not to mention broader and more specialized physics, numerical methods, and engineering journals) so the number of requests is potentially large for the more visible scientists and the probability high that more than one request will arrive at any time (an interesting exercise in Poisson statistics). The review is *not* a random process but there *is* a significant stochastic element at the beginning.

There is also the occasional author – fortunately a very small minority – who simply refuses to participate in the process as a referee. This is not only unfair (well, perhaps not considering what might result in the report) but also irresponsible. Authors take a referee's time, whether they want to be peer-reviewed or not. By the submission they explicitly accept the process. It is a singularly uncollegial act to deny the same journal some effort to provide advice in reciprocity[20]. Some journals explicitly recognize referees. For instance, the AGU and APS have special awards for reviewers who, in the view of the editors of the respective journals, have been particularly noteworthy and conscientious, *Icarus* publishes a list of the referees at the end of the year, and it is not infrequent to see in a *curriculum vitae* a list of journals for which a scientist has been asked to serve as referee. It is a special pleasure for an editor when a reviewer not only enthusiastically accepts the task but, after a careful report, is happy for having been asked. It *is* an honor to be asked to serve as a referee: those who receive a request for a report are asked based on their expertise and publications, not simply because they happen to be on a list or in the bibliography, that's only the start to jog an editor's mind and to keep the pool of reviewers constantly renewing and open.

A potential reviewer should immediately indicate possible conflicts of interest, for example competing research unknown to the editors, that might compromise the process. It also happens that, in refusing a request, a contact may explain

---

[19]This has been a particular nuisance, or worse, in the past few years. Some addresses have been flagged without notification. For example, mindlessly applying the rule that uppercase letters with numbers, or words such as "urgent", may indicate spam or virus-laden mail, some sites have blocked the journal addresses (*e.g.* AA), necessitating workarounds. The database a journal maintains of email addresses is only as up to date as the last contact and even those entries may contain errors, from changes of domains or changes of names. These all slow the process.

[20]The referee's history is also indicated in some journal databases, *e.g.* how many times she or he has been contacted during some calendar period and whether the reviewer accepted the request or not. This is also an important reason why it is a requirement that the submitting author signal whether the paper had been previously submitted elsewhere. This helps avoid multiple reviews for separate journals by the same reader, something that really happens and can lead to interesting results.

biasing predispositions regarding related work by the authors. These notifications are taken quite seriously and are basic examples of ethical behavior.

Assuming all goes well, a paper is placed at the first request, but there may still be a delay waiting for the paper to be retrieved[21]. This is flagged, after some preset waiting period, by the database itself based on experience of the editorial office. We allow 10 days before a flag is raised. A few reminders at this stage may show that the review cannot continue and a new round of refereeing requests begins. Otherwise, when the flag drops and the review is assumed to have started, there is a wait of between 3 to 4 weeks for the first report, for the Letters it is one to three. That is, again, if all goes well. The statistics show that the average reviewer will need one reminder, send before the journal's internal deadline but after the period usually anticipated for the review. Some referees have been known to disappear at this stage, a serious, unexpected problem.

3.3   The Referee Report

The best referee report is one that serves both the authors and the editors. For the authors, it provides an independent, critical reading of the paper, suggestions for improvements or expansions, and flags weak assumptions or sloppy or nebulous presentations. For the journal, it signals where there may be serious problems (for instance, plagiarism, see below), improper citation of – or misrepresentation of – the existing literature (and this abuse goes far beyond issues of credit for discovery, etc.), redundancies (including incremental extension of previously published results of insufficient significance to warrant separate publication), or basic errors. The report should be detailed, grave problems must be flagged and extensively discussed. The report is read by the editor before being sent to the authors, and this includes possible correspondence with the reviewer to clarify points or to request extensions. It is possible that the editor her/himself will add comments that must be addressed in the revisions and responses. If the review is uncivil, or too strong, or the language is poor or ambiguous, it may be edited before being sent out. If it is ambiguous, this is always cleared up before it is forwarded to the authors[22].

The report itself is structured, at least in part, by a set of standard questions to be addressed by the reviewer[23]. These are similar across the spectrum of journals,

---

[21]Even with the best of intentions, there can be delays during the process; as John Lennon wrote, "Life is what happens to you while you're busy making other plans" (*Beautiful Boy*).

[22]The report may also, depending on the journal and editor, be check for possible sexist – or "gendered" – language, in part because the report could be unintentionally offensive (this is one of the cultural shifts from 20 years ago and while it can lead to the occasional awkward construction it has great advantages in preventing misunderstandings). This does not apply to the confidential comments that may precede the report, but even there it may be indicative of problems that need to be discussed before a report is sent.

[23]We have an unusual one in astronomy, whether the objects are denoted according to IAU nomenclature. This is frequently confusing, perhaps even amusing, to those outside the field but the importance of SIMBAD and the enormous importance of the CDS, as Laurent Cambrésy discusses in his chapter, makes this an essential check of the manuscript. Think of what chaos can be created by changing the name of an object or using a nonstandard format for a catalog.

following the historical lines I outlined earlier, and are to be explicated, in detail, in the review. The best reviews can extend for many pages, not that length alone is a criterion. If the depth of reading appears superficial, it may be sent out but with a comment by the editor; in the case of a negative review this may lead to an immediate offer of a second referee (if the correspondence with the referee has not produced a revised version). Since authors may become referees, it is appropriate to insert a few additional comments.

A central issue, "why should this paper be published", is stated rather baldly but addresses a key issue. If the material is merely incremental – what has been sarcastically called a "minimum publishable unit" – or contains largely uninter-preted data (the sort of paper that, in years past, went to the *A&A Supplements*, now discontinued), it may be deemed inappropriate for publication. A separate question is whether papers in a series can be combined or if the series is worth continuing. For this reason, it may be appropriate to have a single reviewer take more than one such paper (if they arrive in tandem) or to line up the same re-viewer for future submissions, should the report show that this will be useful. It is always a potential problem when different editors handle papers in the same series without coordinating the reviews. There are various levels of revision. For cases of truly minor changes, mainly presentation (*i.e.* tables or figures that require re-structuring), the referee exits the process. If, instead, the changes and suggestions are basic to the science but not especially critical, the paper will still go back to the referee in revision but it is virtually at the end of the process and such second reviews are often a formality. At times, however, even these may take extra time if the authors have not properly addressed the criticisms. This does not mean they have not changed the paper, but ignoring a comment is more than bad form, it is incorrect. The comments are included for a reason, and the reviewer and editor deserve explanations if a critique is deemed irrelevant in the authors' judgment. If the criticisms are particularly serious, perhaps fundamental, then the authors may receive the suggestion that the paper be withdrawn and reworked – not merely rewritten – and resubmitted as a new paper. In that case, the paper will enter the queue at the top and may go to a different scientific editor. If the authors choose instead to continue the review, then a very substantial effort is likely to be required and the review may take as long for the revised version as for the original.

3.4   The Review Process as a Queue

The load of an SE varies not only with the day of the week and the season, but also with the hardest of all variables to predict, the return time for a revised manuscript. This is the least controllable part of the refereeing process and, as Bertout's (A&A)and Abt's (ApJ) studies have shown, accounts for the bulk of the time required before a final decision is made on a paper. The median time

---

It may seem an annoyance but without this standardization it may be impossible for work to be checked or may even lead to incorrect search results or citation problems. We ae hardly the only field to have this problem, think of chemistry, geology, and biology.

is relatively short (see Bertout's discussion) but the process is far from Poisson and there is a "fat tail" distribution in the waiting time. The refereeing process is actually a queue, for A&A it is a multiple server, Markovian process[24]. That means there are several servers (a word that carries various connotations, in this case it is the editor) with the papers being arriving units with some temporal distribution on first entry. The service time, the time required to exit the queue, is much more complicated than the usual theoretical problem. There are several components to the waiting time:

1. Arrival rate: papers enter the queue when received at the editorial office through electronic submission, regardless of the journal. The rate is, however, not constant. There are trends through a year (for A&A in the period 2006–2008 – this is discussed in more depth in Bertout's chapter – this varies by up to 50%, with minima generally in Jan./Feb. and Aug./Sep.)[25]. This has a ripple effect through the review process since it determines the load of an SE.

2. Assignment to an SE: almost a deterministic process but may depend on the current load of the appropriate editors;

3. The initial search for a referee: normally of order $n_c \Delta t_c$, where Δt_c is a mean waiting time for a response, allowing for a second contact if appropriate, hence n_c, the number of contacts, includes multiple requests to single reviewers;

4. retrieval time: this is usually short, but can be comparable to the initial search, $1 \leq \Delta t_r \leq 15$ days. This can, and likely will, be reduced and depends on the part of the journal (i.e. Letters or Main Journal). It is a carefully checked step and reminder letters and other contacts are used to prod the start;

5. The review time for first pass: Here a few $\leq \Delta t_0 \leq \Delta_{max}$ where the upper limit is flexible but generally 4 weeks. This depends on criteria set by the individual journals and sections, from a few weeks to a few months (not in

---

[24]I am introducing this way of viewing the process for two reasons. Astronomers know little about queues beyond banks, supermarkets, and traffic jams and it is a very lovely research area that could be exploited in astrophysics. The other reason, more germane, is that it provides a synthetic tool for understanding any process that depends on flows within a service structure.

[25]There are many good sociological reasons for this variation. For instance, observatories have fixed dates for proposal submission, as do many granting agencies. These often cause authors to push the submission dates for papers to demonstrate their productivity. Periodic meetings – the IAU General Assembly, or the AAS winter and summer meetings, or the AGU winter meeting – for instance, can have the opposite effect, temporarily removing both authors and referees from the pool due to the main meeting and attendant symposia. The start of the academic year, and its end, are also perturbations for both parties. There are, of course, the particular months so well known in Europe of July and August when the world beats a path to vacation, even astronomers, and there is always the start of the new year.

astrophysical journals but common in mathematics and history/philosophy of science, and even informatics);

6. The processing time for the first report: this is complex, it may include correspondence between the editor and referee on specific questions but in general it is also short, usually only a few days);

7. The waiting time for first revision: $\Delta t_1 \geq 1$ day: There is an effective upper bound that depends on the journal, when a revision has been awaited for so long that it is effectively a new review because the referee (and/or the field) have changed. The paper may be so radically altered that it's a virtually new submission if the return time is too long. These files are closed by the editor, who periodically culls the papers in this stage depending on date of last action but it can be reopened. This is not a rejection *per se* but amounts to one *de facto*. The distribution of Δt_1 is dominated by correlations and fat tail distributions being closer to a power law than a Gaussian;

8. The return time to the referee: This depends strongly on the time of year and may be longer than the first contact simply because there is no "right of refusal" at this stage, if the reviewer is not available the manuscript sits in a holding pattern);

9. Report on the revised version, Δt_2, which generally has a similar form but truncated limits relative to Δt_0. One note: a revised paper may require as much, perhaps even more, time to referee than a first submission, the authors can help significantly by carefully detailing their revisions in the manuscript and thoughtfully and extensively responding to the report (we request keying to the relevant items) in their cover letter (which is forwarded to the referee along with the manuscript). One reason, alas, for a delay is that the responses are not always in accord with the text of the changes, or the reasons included in the cover letter are so important that they should be included in the paper;

10. This process is iterated, usually at least once but perhaps more times, see below for a discussion of how impasses are resolved.

There is a parallel queue for *Letters*. Most journals, if they have such a section, create an independent editorial structure to oversee the process, including SEs and a separate referee pool. This review procedure can actually set the journal in competition with itself if the possibility exists to transfer a paper, at the level of the editors-in-chief, from one part to another. At A&A, a radically different structure has evolved with the SEs overseeing both parts, thus allowing seamless transfer between the two. Such decisions are made according to well-defined, although flexible, criteria.

The most difficult step is the wait for the return of the first revision. The timing is important and uncontrollable. The load of an editor can vary but for A&A it consists of some number of manuscripts to be placed, some number awaiting a referee, and those in active review, and can be between 100 and 200 at any time. Some fraction will be in first review, some in second, some in third, some in second refereeing, so you get the picture. Some further number will have been recommended for acceptance and are on hold, some will be with the language editors, others awaiting publication. That the process takes a long time can, I hope, now be easily understood.

3.5 Second Referee

This is left open-ended because there is a possible branching factor. Let me explain. Some journals use two or even three reviewers and, based on a weighting scheme, determine if the paper is to be refused at first review. If a revision is allowed, it may still lead to a rejection, or it may be rejected up front at the revision stage by the editor. In single-referee review processes, should the editor determine that an impasse has occurred, or if the referee signals an unwillingness to continue, the authors may be permitted to request a second reviewer – with justification and appropriate revision of the paper before submission. It can also happen that there is clearly no point because the referee has simply refused to reconsider the manuscript after the first review. The second reviewer can be independent or "informed", the first being a completely new review of an obviously revised paper (at least with knowledge that the paper was already through one review cycle) while the second means the responses – the history of the review cycle – has been supplied. There's a sort of third option, sometimes requested by the second referee *or* offered by the editor: at the end of the review of the revised paper, once the report has been received at the journal, the referee may get to see the original comments as a calibration of her/his report. This is, of course, one of the reasons the editor's oversight is so important – the editor sees all correspondence and can judge whether the report of the first reviewer is really to be ignored, or that of the second, or whether there are common points that should be signaled in the cover letter to the authors.

4 Particular Issues

4.1 The Availability of Information: From *Jahrsberichte* to the *Astronomical Data System (ADS)*

The world is awash in scientific publications. There are hundreds of conferences, each of which produces proceedings, and hundreds of thousands of pages published each year. To say it is folly to follow the literature at such a detailed level is an understatement (more on this when we discuss citations). But since the days of the *Phil Trans.*, there have been aids for the bewildered and overwhelmed. These are the abstracts published by dedicated reviewers of the journals since the 18th century. A particularly noteworthy example is the review journal *Astronomischer Jahrsberichte*, which published reviews of the astronomical literature for many decades, which was a review abstracting service, still extant in the mathematics community as *Mathematical Reviews* assigned papers for review to a large pool of "post publication referees". These were/are more than abstracts; they are usually brief (or not so brief) critiques of the paper. The *Astronomy and Astrophysics Abstracts* was different. As the forerunner of the Astronomical Data System, it was a compilation of the abstracts of papers. Its principal advantage over merely browsing the shelves was an extensive taxonomic system that placed cross-referenced keyword codes alongside the abstract to render the paper more accessible (as Abt did for the *ApJ*). This was the underlying informatics architecture adopted by the ADS, that permits searches by keyword – not merely title and author – but with the addition possible only with the development of sophisticated database manipulation, to link papers at many search levels, also linking the reference lists (as done also by *Science Citation Index* and related databases) but also connecting to the paper itself and to data and other material relevant to the study. This is covered further in Laurent Cambresy's chapter on the workings of the Centre de Données Astronomiques de Strasbourg (CDS) in general and *SIMBAD* in particular, so you are referred to that discussion. The journals still maintain their own categories and keywords, subject to revision over time by editorial review, and the authors can choose the subsections for those journals that explicitly have them (and, collaterally, to state a preference for the choice of scientific editor).

4.2   On Anonymity

The referee process appears lopsided. The referee knows the author(s), while the author(s) generally do not know who the referee is. You know from grant and observing time evaluations, and job applications, that this is very discomforting. But it has some value in keeping the process impersonal since the mediation, and moderation of the discourse, is managed by the SE[26]. Some journals use double-blind reviewing, where both sides are anonymous. This is used frequently outside of the sciences so let me explain how it works (since it is possible you have never seen the process in, say, history).

The main difference between the editorial role I have been describing and the editor of, say, a journal in the history of science is to insure complete anonymity. This is why you see such stilted language in some disciplines, the impersonal is used to hide the author's identity and parts of the text may even be "censored" (read "blacked out") to insure that self-citation does not reveal anything (*e.g.* instead of saying "as we showed in Blop *et al.* (2008)" this would read "as Blop *et al.* (2008) have shown", a minor difference in wording but one that signals a very different mindset. When in some areas of research the authors are many and the individual products are relatively few, this works. We, on the other hand, are more accustomed to building a context by citing previous work by ourselves as well as others and part of the value of referencing is how it can – used assiduously –

---

[26] As a procedural note, the scientific journals have, by and large, adopted the "Nobel Prize rule", releasing their referee reports for scholarly use after 50 years. Some journals make this an explicit agreement, signed by the authors and known to the referees. For others it is implicit.

keep a paper short but still inclusive. In grant reviews this is obviously impossible since the track record of an individual or group enters the weighting scheme for estimating the likelihood of future success[27]. It is not, in astrophysics, all that hard to narrow the list of possible authors or possible reviewers given the citations. So in a better world, it might be reasonable to go this way but in practice it is really the editors' role that makes this less necessary.

The opposite choice would be completely open review. This is reasonable and democratic but not without possible abuses. The public posting of reviews of papers, as some have suggested for open access, is based on a market analogy that can seriously damage a paper and/or its authors and might well lead to a mere degeneration toward the mediocre. By this I mean that if both parties are airing their views in public, both parties may be cautious for careerist reasons – their own and/or their students and/or collaborators – since the fear of revenge is a very human feeling and not easily overcome. Moreover, in a litigious society, a feature of almost all western countries, the consequences may be even more serious[28].

Open reviewing is, therefore, now left as a choice for the referee. Correspondence are requested for the journal files, and the information of the editor in charge of the review, which may be critically important at acceptance. This can be very useful to the referee, since it makes the use of direct discourse far simpler. But it can – fortunately very rarely – lead to "referee abuse", with the authors making the reviewer a *de facto* collaborator without even an acknowledgment (yes, it happens). But again, it is rare and editors often query the reviewer who has selected this option to be certain (the choice of waiving anonymity can actually be an error)[29].

4.3   Referee Ethics: The Norm and the Breech

On accepting the task of reviewing a manuscript, the referee accepts the implicit (sometimes *explicit*) rules governing the review. In the days before electronic

---

[27]This is an example of the Matthew Principle, explicated by Robert Merton (1968) and elaborated by generations of sociologists of science but it has its value, along with bibliometric studies. Those who have been successful, or have been in the forefront of a research program, are often in a better position to make future advances or make reliable statements when presenting their results. An illustrative application of this to citations is found in Brown (2004).

[28]To illustrate this point, a simple review of the costs of scientific journals published in *Physics Today* in 1983 led to a more than decade-long international lawsuit by a presumably aggrieved publisher resulting in enormous costs, personal tragedy, and an inconclusive and generally unsatisfactory end.

[29]At times the editor may recommend, for the first pass, an anonymous review to facilitate the process, leaving open the option of full disclosure at a subsequent stage. This is done in consultation with the referee. There is one other alternative used by the *AGU* journals. Referees are openly acknowledged, along with the indication of the responsible editor, at the end of the *published* paper. This is by agreement with the reviewers, some of whom nonetheless prefer anonymity. But these papers receive more than one referee and, in the case of *Reviews of Geophysics*, often more than one depending on the interdisciplinarity of the article. The editors are also solicited for nominations for special recognition of notable reviewers by the society at the end of the year, published in *EOS*.

preprint servers, the reviewer was in a uniquely privileged position, as the person who would see and critique a work before it was offered to the community at large. Even with the alternative means now available to permit access to the manuscript, the referee is ethically bound to behave with the same discretion and honesty, not revealing the contents of the work to colleagues and considering the manuscript a confidential communication. She or he must not make any use of the material, exactly as in the review of a proposal for observing time or a grant, or in any manner deliberately interfere with the review process. To violate this is more than a serious issue: it is a capital offense against the only thing that keeps the scientific enterprise functioning in the open, it is a breech of trust. The ethic of scientists is to be open, and even if the reviewer is anonymous it is perhaps the most important role of the editor to insure that there is no transgression. If a conflict of interest arises during the review, it is the obligation of the reviewer to immediately notify the journal. If this is clear at the start (sometimes it is not obvious from the abstract and only becomes evident once the actual manuscript is in hand) to not send a notification to the editor is dishonest. For this reason, among others, it is best to at least quickly give a quick pass through the paper to identify possible problems. Most of the time, in fact the overwhelming majority of cases, this is exactly how it happens and the problems (which are few in any year for any editor) can be dealt with quickly. But there are some who, by ignorance or design, seek to delay or even kill a paper for their own benefit. I repeat, this is a rare event, but "rare" is – alas – not "unknown". This does indescribable damage at many levels. It weakens the faith in the efficiency and transparency of the review process. It reinforces the prejudice that the publication of a scientific paper is a sort of struggle against the Dark Side. A few examples are the reminiscences of Parker (1997), the history of Einstein's dealings with *Phys. Rev.* by Kennerfick (2005), and the particularly caustic remarks by Eggen (1993), and also Cole *et al.* (1981) for the view from sociologists. There are places reserved in *Inferno* for this sort of person.

4.4   Citation: The Editorial Side of the Issue

The footnote was one of the most important inventions in scholarship, but it was preceded by the gloss – the marginal comments inserted by readers of manuscripts in the centuries before printing – that often not only expanded on the text but indicated the sources in a way not done in the text. It was not unusual in the Middle Ages for texts to contain pieces lifted from other works without citation and it is still one of the tasks of scholars to note the sources of the material (which can often lead to new historical insights). This was no mere token of authorship, but was instead a way of contextualizing the material. A citation meant, and means, the full source, with all its individual peculiarities and content, can be traced back without the need for the author(s) to duplicate the information. There is, of course, the other reason, that a living author explicitly acknowledges an intellectual dependence and debt with a citation. It also means that the work has been read and understood (well, at least read). One not-so-humorous definition of a *classic* is a work that is frequently cited but rarely read (see *e.g.* Evans 2008). How many of you, for example, have gone back to the original papers on general relativity, quantum mechanics, stellar structure? Even if these have now passed into history, suffering the fate of "obliteration" – another phenomenon identified by Robert Merton in the 1940s, when a work becomes common intellectual property and the original work no longer needs citation – it is nonetheless assumed that cited it has been read! The same is true for any more modern paper.

Review papers have a special place in this discussion. Its scope is often to assemble and order a vast, dispersed literature. But in the process, a single review paper can effectively "kill" the literature it attempts to synthesize. The most cited papers are, generally, those in *Annual Reviews* and *Reviews of Modern Physics*. Of course, one reason is their singular utility, points of reference from which a study can depart without recreating the whole of the earlier work in a field. But at the same time, the review can be cited (the proverbial "see ... and references therein") as the unique precedent for any study without taking the time to read either it or any of the literature on which it is based. In an important study, Brown (2004) discusses the diffusion of such citations in the literature based on internet searches).

Bibliographic "purity" is very hard to achieve but it is as important as nomenclature. The error rate in bibliographic databases is improving but it cannot reach perfection so imperative that authors do their utmost to play their part. We cannot rely on the librarians and it just makes their life harder when we producers of the literature are sloppy. A wrong citation isn't just a matter of incorrect formatting. It can create a work that doesn't exist. A phantom paper is a permanent problem. I give a simple example. Last year I was checking the citation of a classic paper, the founding paper on the Kelvin-Helmholtz instability. The text and bibliography referred to it as Helmholtz & Monats. But the actual paper is by Helmholtz alone so I checked the ADS and then started a sort of traceback through the physics journals. The paper is a phantom that's propagated through the literature. It *should* be Helmholtz alone – *Monats* is actually the abbreviation for *Monatsblatt* (Monthly). As a result, there is now a permanent record of a well cited author who was born and died in a single year and produced only one notable paper in his short life. To coin a phase, I will refer to this as the "Lt. Kije" phenomenon[30].

4.5 Series

How do you decide whether a paper should be published singly or as a series? As publication and citation counts become progressively more important for careers, so has the proliferation of series. These may be implicit – very similar,

---

[30]For those who love music, the suite by Prokofiev by this same name will be familiar. It is based on the story by Yuri Tynyanov (born 1894, died 1943) about a mythical Russian army officer created by a company of soldiers whose birth, life, and death are fabricated to fool bureaucrats.

cross-referenced papers published in tandem or successively – or explicit, with division into labeled "parts". This is not new, of course, since whole books have been constructed from some of the historically more famous examples. The Dover reprint volume *Physical Processes in Gaeous Nebulae*, edited by Menzel in the 1960s, is largely composed of the *ApJ* series of the same title that ran to 15 parts from 1937 to 1945 (Aller 1999). Chandrasekhar's series "On the Radiative Equilibrium of a Stellar Atmosphere" had 24 parts in the *ApJ* from 1944 through 1948 and was finally condensed in his monograph *Radiative Transfer* in 1950. The series "Wavelength Dependence of Polarization" appeared in 40 parts in *AJ* from 1960 through 1980, many dealing with single objects, which spawned volumes in the Arizona series of astrophysical review volumes. The alternative would have been to give each a separate title, perhaps less obviously sequentially connected, and use more space in each recontextualizing the work.

A few words of advice. First, the decision should be based on two things: readability and scope of the work. If too many related but disparate topics are presented in a single paper, the reader may well lose the thread. It may be better to split the paper along some natural lines, but this should be done at the start. In such cases, the parts should be "natural" units that really follow each other and build a result. This does not mean, for instance, that multiple objects or case studies necessarily should be – or even deserve to be – separated. On the contrary, the case may be stronger if statistically meaningful samples, or related objects, are presented together.

The decision should be scientific, not stylistic. It can be embarrassing to have a Part 1 published alone, and if dealt with by different editors and/or referees it can happen that one or more of the subsequent papers do not make it through the process or is combined, leading to a peculiar sequencing (*i.e.* when, say, Paper 4 is accepted but not Paper 3). Such things happen. It is, therefore, vitally important that the papers do not "refer forward" – that is, one paper does not leave *essential* details to be explained in a future submission. A review cannot be based on a promise of details of an essential item or draw conclusions based on results that have not yet been either submitted or completed. This is true for *any* paper: material cited as "in preparation" is unacceptable if it is at all important for understanding and judging the results in a manuscript. It may limit the choice of publication venue since some journals are reluctant to continue the numbering of papers in a series for which some have appeared in other journals. It may also lock the series into review by a small pool of referees and will usually be assigned to the same editor (although it happens that sometimes several editors handle the series but may consult among themselves).

Although little studied, Schultz (2010) provides a snapshot from the editorial side (as editor in chief of *Monthly Weather Reviews*, one of the American Meteorological Society journals). If submitted together, for instance, as Parts 1 and 2, his study showed an acceptance rate essentially the same as a single paper. If submitted serially with some time between the arrival of the two parts (*e.g.* if one has already been published), the second has a higher than average acceptance rate. This should not be misunderstood as advice on how to manipulate the review process. Quite the contrary. The higher acceptance rate (about 3/4 compared to 2/3 for the average) is because the multiple parts are linked well and are logically sequential units. When the papers are submitted together, there is also the chance that – sent to a single referee – they will end up combined (although Schultz found, from a ten year sample, that the acceptance rate for the two together was the same as single papers). His final advice is to "write manuscripts that are sensibly [sic] independent of each other, make minimal reference to unsubmitted manuscripts, and have sufficient and substantiated scientific content within each manuscript", all common sense notions but more often ignored than followed.

4.6   Multiple Submissions

This can be briefly stated: the submission of a paper to more than one journal is unethical and prohibited. Period. Such a transgression can lead to an author being banned, perhaps permanently.

4.7   Alterations to the Paper Following Acceptance

This can be an ethical issue if the changes are not indicated when the paper is returned to the publisher after proofreading. Any changes in the content (as opposed to the form) of the paper *must* be indicated as a "note added in proof" and may result in the paper going back to the referee. It is not enough to say that the substance/conclusions are unchanged, at times this is not true; a detailed explanation must accompany the revised manuscript. Not respecting this can result in harsh penalties, analogous to those reserved for cases of plagiarism.

4.8   Plagiarism: Intellectual Property and Authorship

To explain the problem and disease – not too strong a word – of plagiarism requires a context. The notion of intellectual property is fundamental to the scholarly enterprise and as old as the scientific societies and journals. At its core lies a very human feature of what we do, wanting *our* contributions be recognized by our peers as original, perhaps even valuable efforts in the advancement of science. This is very different from ownership: what we produce, think, and communicate is given openly and freely to the community at large. It hardly even hints at a capitalist idea of property leased or rights sold. The idea of copyright was born in the 18th and 19th century as a way of protecting the expropriation, in part wholesale, of an individual's work (see Rose 1993 for a superb history). Our compensation, if there is any way of explaining it in such terms, is that we have attached to our names (in a singular or collective sense) the air of reliability and novelty, of creativity and knowledge. These are abstractions when so categorized but they are the very personal reason most of us become scientists, to participate as equals in an international community in a historical enterprise. We all know that the literature is ultimately the basis of what we do. It is possible that some aspects of physical and mathematical research could be conducted on a desert island from first principles but there are few imaginable examples and they have, in fact, never occurred. Music and art can spring spontaneously, even calculating prodigies exist, but this is not the same as what we do. All of us have passed through a formative period in which we absorbed the cumulative results of the past centuries, the models, theorems, and empirical results of the preceding generations. Some of these have names attached, not always assigned properly by history but indicating, in some sense, the recognition that individuals have done, thought, or discovered something. Others have become so commonplace and essential that we no longer even note the authors, they may now seem obvious to us, but they were still human products.

It is in this light that the act of expropriating someone else's ideas, results, or words – plagiarism – is an offense, to those parties affected and to the scientific community as a whole[31].

As I said above, the results of a scientific paper are common property but not its authorship. How a result is arrived at, understood, and explained can be as important as the result itself. Alone, a fact is nothing. Set properly in a broader picture it becomes a resource. If something cannot be understood by others it is worthless. We communicate in structures, words and equations and graphics, the results of our studies and these are particular to the source.

The concept of authorship is not just for "credit", a token to be used in job and grant applications, but that also plays a role. We live in a world that also judges what we're doing for the purpose of awarding (never forget the various nuances of that word) positions, observing time, support. Every time you apply for time at an observatory, beam time at a large national or international facility, a fellowship, or a position, a research grant, or even for a PhD program, individual contributions are the basis of the judgments. They are also the basis for choosing those who will be the judges. True, this is an idealization of a process but there is an idealistic side to the "scientific society", it *must* assume that those who are involved in all levels are fair and honest. For the cheater, to expropriate something gets ever easier. Previous generations had to work a little harder to be successful plagiarists. Before there were databases of published articles, or electronic reproduction, it was necessary to go to a library, or at least open a journal or book, and manually copy the material. The act required such effort that only the dedicated were willing to lift something. With a smaller community and fewer publications, it might even

---

[31] As an illustration of the point, as this chapter was being completed the German defense minister, Karl-Theodor zu Guttenberg, was forced to resign and renounce his PhD from the University of Bayreuth as a result of plagiarizing passages from other scholars that produced a protest by the academic community against treating this as a minor episode. As quoted in the NY Times (2011 Mar.1, reported by Judy Dempsey), "... last weekend more than 20 000 scholars from Germany and other parts of Europe sent an open letter to the Chancellery saying that Mrs. Merkel's continuing support of Mr. Guttenberg was a "mockery" of all those who "contribute to scientific advancement in an honest manner." "If the protection of ideas is no longer an important value in our society, then we are gambling away our future", the statement said. "We do not expect gratitude for our scientific work, but we do demand respect. The scientific community is suffering as a result of the treatment of the Guttenberg case as a trivial offense. As is Germany's credibility"". You see, there can be very serious consequences.

have been harder to get away with it. It was a deliberate, egotistic, conscious act. Technology allows anyone to cut and paste almost anything. This, along with the "democracy" of the web, seems to blur lines between the work of a colleague and yours. If something was said better in a published paper, why not simply use that yourself to explain something or describe something. If a figure serves your purpose, why not simply show it in a talk? Nobody will care because it is not the central point of what *you* are doing or saying, it is background.

But it is not. Fitting a curve to others data without citing their source is plagiarism. Changing the data to fit the model, or select those that fit and discard the rest, is fraud. And the opposite is just as true: showing a result, any result, without proper attribution and explanation of the criteria for the comparison is simply wrong and inadmissible.

One of the roles of the referee-editorial process is to help guard against such abuses, committed consciously or unconsciously. The ethics of authors is taken for granted, it is not the norm to assume that a paper is fraudulent or is misrepresenting the work (*e.g.* LaFollett 1996).

There is one additional form of plagiarism that, while not seeming serious, is unethical, the direct use of the text of the referee report. The referee may make suggestions on specific wording. In such cases, a properly phrased acknowledgment is sufficient to indicate that the authors have received direct "editorial" help from the reviewer. In other cases, however, the text of the report may contain a critique that is *not* editorial but intended as a critique to be addressed by the authors with textual revisions and further explanations (in the cover letter that accompanies the revised manuscript). This is *not* to be simply copied into the text of the revision, yet there are instances of such appropriation. This is laziness on the authors' part but it also does not address the critique, but merely acquiesces to the reviewer in the hope that the change will make the paper acceptable. In fact, it should do just the opposite since it shows that the authors are doing "whatever is necessary to satisfy the referee". To an editor this is absurd. It shows an ignorance of the point of the refereeing process and may even produce unresolved logical contradictions in the paper.

A recent development is the employment by a number of journals of automated commercial software to check for plagiarism. Functionally similar to search engines, they provide a "matching statistic" that signals potential problem cases. This *never* replaces the human evaluation but it is efficient and usually reliable as a first step, often being used because the editor notices something even before the paper is sent to a referee[32].

4.9   Ghost and "Honorary" Authorship

It has been said that "on election day, the dead rise" and this also happens with journals. The phenomenon of ghost, or *honorary*, authorship is yet another strain

---

[32]This is well described in an editorial and several opinion pieces in *Nature* (2010, *Nature*, 466, 159 Jul. 8, see also Titus & Bosch (2010) and Koocher & Kieth-Spiegel (2010).

on the ethical cords of scholarship. The stories are countless and, some, extremely sad[33]. The honorary authorship may lengthen a publication list but it is dishonest, both for those included and those who include them. It is sometimes done believing that it will ease the review process by seemingly giving a "stamp of approval" by senior and/or well known scientists to the work of a junior or less well known colleague. It may, however, have just the opposite effect. Should any of the "ghosts" discover their names in an author list of a paper they have not previously seen, e.g. after being posted on astro-ph, they may write directly to the journal, initiating an investigation and possibly leading to sanctions against the offenders. Some journals require an explicit statement of responsibility – who did what in the research and the writing (e.g. Nature), while others take this as implicit – as well as "conflict of interest" statements. In very large collaborations[34], there are usually internal agreements that require participants to be actively involved in the work in some way during a period of time. It is a universal editorial "given" that the paper has been reviewed, and approved, by all concerned. There have been cases, fortunately not in astronomy, when a paper has been shown to contain fabricated results, in those cases embarrassed ghost co-authors have been held responsible along with the offenders, often doubly so because they are forced to use the explanation that they did not even know what was in the paper and, thus, cannot be held liable for the falsifications. Think about what this says about the scientist making such a defense (noting, mainly from very high-impact journals, some of the recent scandals in material sciences, biomedical sciences, and famous cases in paleontology; it is enough to recall the case of recent bubble "cold fusion").

4.10   The Role of Acknowledgements

In light of the last section, let us finally turn to the last part of a published paper. The role of acknowledgments is to make public the indebtedness of the authors to those whose counsel and help does not merit authorship but who, nonetheless, contributed in some way to the final paper. This is no mere social token. It makes clear the role of those who have been involved in the background. This includes the referee, even anonymous (it is not ridiculous to thank an anonymous reviewer, the referee and editor know who it is and it makes clear to the rest of

---

[33]One, however, is actually funny. The author list of Alpher, Bethe, and Gamow's paper on cosmic nucleosynthesis was the product of Gamow's peculiar sense of humor but Bethe was actually aware of the work and agreed to the $\alpha\beta\gamma$ ordering.

[34]This is a growing feature of astronomical research that has been commonplace in, for instance, particle physics for decades: the Virgo and LIGO collaborations for gravitational wave observations numbers in the hundreds, similarly for the Cherenkov telescopes MAGIC, H.E.S.S., and VERITAS, large space projects such as Fermi, and surveys like the Sloan Digital Sky Survey and 2MASS. Of course not everyone in groups of this size contributed at every stage, but there are internal refereeing groups, sites on which the preprints are posted as drafts for internal comments, and a host of other mechanisms for insuring that those involved take appropriate responsibility for the contents of the papers issuing from the projects. Often lead authors are identified as the contact points so they are visible even when an invariant order of authorship has been worked out by negotiations within the collaborations.

the community that the refereeing process has been effective). It is particularly uncivil and uncollegial to *not* thank a referee when the review has been open. This section also provides the antidote to the ghost author attribution, an appropriate acknowledgment suffices.

There is now another, unfortunately more bureaucratic reason to pay attention to acknowledgments. Increasingly, in the face of threatened or decreasing budgets, institutions and funding agencies seek to justify their existence, authors are under pressure to assist. Just as plagiarism sniffers can detect miscreants, the same technique is employed to count "citations" to facilities and grants. Many institutions have specific text that must be inserted when time is granted, whether a supercomputer center of observatory. Others, such as the CDS and ADS, use this to document their centrality in modern astrophysical research. You are advised to always check with these entities regarding the wording[35].

5 Concluding Remarks

It is clear from the history of science that the publication process, based on external peer review and scientific editorial oversight, is neither pure nor perfect. But it has evolved and can be improved. It is as unending an experiment as science itself.

My sincere thanks to Claude Bertout and Helmut Abt for their wise guidance and for innumerable, golden discussions and exchanges over many years, to Thierry Forveille, Joli Adams, and Chris Sterken for their comments on this chapter and for our discussions over many years, and to my fellow editors at A&A in this amazing enterprise for their scientific and editorial collaboration. I thank Jason Aufdenberg and Pierre Henri for their detailed critiques of the text, and Johannes Andersen, Daniele Galli, Enore Guadagnini, Jordi José, Rob Kennicutt, Ted LaRosa, Leon Lucy, John Mariska, Georges Meynet, Jennifer Martin, Pascale Monier, Birgitta Nordström, Jan Palouš, Francsco Pegoraro, Pier Giorgio Prada Moroni, Paolo Rossi, Greg Schwarz, Luigi Stella, Ethan Vishniac, and Glenn Wahlgren for invaluable discussions over the last decade on these and related topics, I also want to thank the thousands of referees and authors with whom I've had the pleasure of corresponding over the years for many enlightening offline discussions.

Dedication

This chapter is dedicated to two companions: Cody (b. Three Rivers, Michigan, 14 May 1994 – d. Blankenberge, 19 May 2008) and Quillo (d. Pisa 3 Oct. 2010).

References

Abt, H.A. (ed.), 1995, American Astronomical Society Centennial Issue of the Astrophysical Journal (Chicago: Univ. of Chicago Press)

---

[35]Some of the larger projects, the SDSS for instance, have a laundry list of institutions that may seem useless waste of space but the statistics compiled by such acknowledgments can mean the survival of a project. It illustrates, sadly, the increasingly intrusive role of quantification of "research productivity".

Alberts, B., Hanson, B., & Kelner, K.L., 2008, Science, 321, 15: "Reviewing peer review (editorial)"

Aller, L.H., 1999, ApJ, 525, 265: "Menzel's "Physical Processes in Gaseous Nebulae"", commentary in AAS Centennial Volume

Anderson, R., 1993, Notes Rec. Roy. Soc. London, 47, 243: "The referees' assessment of Faraday's Electromagnetic Induction Paper of 1834"

Batchelor, G.K., 1981, J. Fluid Mech., 106, 1: "Preoccupations of a journal editor", a must read for anyone wanting a deep reflective assessment of scientific publishing

Bluhm, R.K., 1960, Notes and Records Royal Soc., 15, 183: "Henry Oldenburg, F.R.S. (c. 1615–1677)"

Brown, C., 2004, Scientometrics, 60, 25: "The Mathew Effect of the *Annual Reviews* an the flow of scientific communication through the World Wide Web"

Eggen, O., 1993, ARA&A, 31, 1:" Notes from a Life in the Dark" (this is where Eggen lets it all out at the referees and editors)

Evans, J.A., 2008, Science, 321, 395: "Narrowing of science scholarship". This is one of the most important studies of citations and the evolution of the "general culture" of the literature, and one that has received considerable attention (by a historian of ancient astronomy, so one of wide culture)

Giles, J., 2006, Nature, on checking of results, "The trouble with replication"

Hartman, P., 1994, *A Memoir on the Physical Review: A history of the first hundred years* (NY: AIP Press)

Hoffmann, D., 2005, Ann. der Physik, 17, 273: " "... you can't say to anyone to their face: your paper is rubbish". Max Planck as Editor of the Annalen der Physik"

Jackson, D., & Launder, B., 2007, Ann. Rev. Fluid Mech., 39, 19: "Osborne Reynolds and the publication of his papers on turbulent flow"

Kennerfick, D., 2005, Phys. Today, (Sep.) 34: "Einstein *versus* the Physical Review": this is a rare look into the archives at the process behind refereeing

Koocher, G., & Kieth-Spiegel, P., 2010, Nature, 466, 438: "Peers nip misconduct in the bud"

LaFollette, M.C., 1996, *Stealing into Print: Fraud, Plagiarism, and Misconduct in Scientific Publishing* (Berkeley: Univ. of California Press)

McClelland III, J.E., 2003, Trans. Amer. Phil. Soc., 93(2), 1: "Specialist Control: The Publications Committee of the Académie Royale des Sciences (Paris) 1700–1793": The most complete study to date on the origin of the refereeing system in the 18th century

Meadows, A.J., 2008, *Science and Controversy: A Biography of Sir Norman Lockyer, Founder Editor of Nature*, 2nd Ed. (London: Macmillan)

Merton, R.K., 1968, Science, 159, 56: "The Matthew Effect"

Parker, E.N., 1997, EOS, 78, 391: a personal account, "The martial art of scientific publication"

Pyenson, L., 2005, Ann. der Physik, 17, 176: "Physical sense in relativity: Max Planck edits the Annalen der Physik, 19061918"

Reingold, N., & Reingold, I.H., (eds.), 1981, *Science in America: A Documentary History, 1900-1939* (Chicago: University of Chicago Press)

Rose, M., 1993, *Authors and Owners: The Invention of Copyright* (Cambridge, MA: Harvard Univ. Press)

Schultz, D.M., 2010, Scientometrics, 86, 251: "Rejection rates for multiple-part manuscripts"

Titus, S., & Bosch, X., 2010, Nature, 466, 436: "Tie funding to research integrity"

Wali, K.C., 1991, *Chandra: A Biography of S. Chandrasekhar* (Chicago: Univ. of Chicago Press)

Wright, H., 1966, *Explorer of the Universe, A biography of George Ellery Hale* (NY: Dutton)

Zucker, R.S., 2008, Science, 319, 32: "A peer review how to"

Some Additional Readings:

Abt, H.A., 2000, *unpublished*: a personal note, this set of comments on refereeing was prepared during Abt's term as editor of the *ApJ* but was, alas, unpublished. It is, however, availble from its author

Berendzen, R., 1974, Physics Today, 27(12): 33-39: "Origins of the American Astronomical Society"

Burrell, Q.L., 2005, Scientometrics, 65, 381: This is a particularly interesting paper, "Are sleeping beauties to be expected?" The phenomenon is a paper that, after years of neglect, suddenly becomes "hot". Perhaps the best example I can cite is the famous EPR (Einstein-Podolsky-Rosen) paper on the completeness of quantum mechanics. For decades it was known, and discussed, almost exclusively in the philosophical and historical literature, along with Bell's theorem. Then along came quantum computing and suddenly these became required reading at the leading edge of theoretical physics

Chilton, S., 1999, Academe, (Nov.) 54: this paper is one of the sleeping beauties, intended for a general academic audience (this is the magazine of the American Association of University Professors (AAUP), the group that defined tenure in the North American university): "The Good Reviewer"

Cole, J.R., & Cole, S., 1972, Science, 178, 368: another example of the sociological view of the referee system, following the same path as Merton's studies, "the Ortega effect"

Cole, S., Cole, J.R., & Simon, G.A., 1981, Science, 214, 881: another of the social studies similar to Merton and the Matthew Effect, "Chance and Concensus in Peer Review"

de Solla Price, D., 1975, *Science Since Babylon: Enlarged Edition* (New Haven: Yale Univ. Press) a real classic, one of the first historical-social studies of the growth of science by a leading historian of physical science, see chapter 8: "Diseases of science"

Garfield, E., 1986, Current Contents, 32, 3: one of the classics, frequently cited, by the founder of Science Citation Index, "refereeing and peer review"

Goudsmit, S.A., 1963, Phys. Rev. Lett., 10, 41: "The future of physics publications: a proposal", an editorial at the start of the literature explosion in the 1960s

Greene, M., 2007, Nature, 450, 1165: as the number of authors multiples, and projects become ever more complicated (in fields that previously had been immune to the "big science" team research structure), this paper offers some interesting insights: "the demise of the lone author"

Holmes, F.L., 1987, Isis, 78, 220: a general historical account, "scientific writing and scientific discovery"

Judson, H.F., 1994, JAMA, 272, 92: one I recommend for its view of the process in perhaps the most sensitive field for refereeing, medicine: "structural transformations of sciences and the end of peer review"

Phelan, T.J., 1999, Scientometrics, 45, 117: a guide for the perplexed in bibliometrics, "A compendium of issues for citation analysis"

Skilton, P.F., 2006, Scientometrics, 68, 73: A recent study of the evolution of citations as they propagate through the literature, "A comparison of communal practice: assessing the effect of taken-for-granted-ness on citation practice in scientific communities"

Suppe, F., 1998, Phil. Science, 65, 381: "The Structure of a Scientific Paper" (see also comments by Alan Franklin and Colin Howson); also Lipton, P. ibib, 406: "The best explanation of a scientific paper"

Weller, A.C., 2000, J. Amer. Soc. Inform. Sci., 51, 1328: one of the early reviews, "Editorial peer review for electronic journals: current issues and emerging models"; see also Brown, C. 2001, J. Amer. Soc. Inform. Sci., 52, 187: an example of the process of expansion of the arxiv viewed from outside, "e-volution of preprints for physics and astronomers"

LANGUAGE EDITING AT ASTRONOMY & ASTROPHYSICS

Joli Adams[1]

Abstract. After its founding, A&A moved to a policy that all articles must be written in English. Once this was established, the next step has been to improve the overall quality of the language in the articles with the help of a team of language editors. This article reviews the general advantages of editing the English expression and describes both the aims of this effort and its place in the full publication process.

1 Background

The very first volume of *Astronomy and Astrophysics* contained only three articles in French compared to the 60 in English, and none were published in the other European languages representing the consortium of the countries that founded the Journal in 1969, although there were occasional submissions later in French and German. At first, the authors were responsible for writing understandable English, so a paper's English was looked at only if it was full of errors. Now and again, the Editors suggested changes, as would the referees; but when it was unreadable, it would either be sent back to the authors asking that a colleague revise the language or else sent to someone with excellent English known to the Editors to revise it because the authors had no resource in their institute for the task.

This early arrangement worked because so many European authors had already been writing in English. It is also understandable, since this new journal had already achieved a lot and was still developing its policies and procedures, so that improving the English expression any further had to await other advances, in particular adding sponsors and then moving the editorial operations to a single site, thereby freeing up some of the operational expenses for more than one full-time language editor.

Most papers in A&A have always been written by non-native English speakers, and they have also been read by non-native English speakers over the world, so in the end the main goal has always been to seek clear and correct English when

[1] A&A Language Editor, Observatoire de Paris, 61 Av. de l'Observatoire, 75014 Paris, France

the A&A editors, referees, and now language editors make suggestions. Like all other changes at the Journal, A&A hopes to provide its authors with the best vehicle for presenting their research and ideas in an international forum. Unclear or inefficient English only stands in the way of this goal.

This article covers the background to the role of English in the Journal, which of course includes why English is so important in international science (Sect. 2). Section 3 tries to explain why the quality of the expression, in particular, is important, so it goes into some of the language problems worked on in each paper. Section 4 describes the process of language editing within the Journal itself in more practical terms. Following some suggestions for other resources that authors can use before and after submitting papers to the Journal, this chapter ends with a brief concluding statement.

2   Why Move to Only English at A&A

In 2001, the A&A Board made official an earlier policy to publish only in English at A&A. This decision was in line with the situation in other scientific journals and with all the other changes made at the Journal over the years to improve its effectiveness as a forum for European science in the world. The attempt to improve the quality of the English began at about the same time, perhaps, and is one aspect of the more general improvements and only one of the many reasons that the Journal's impact has been rising.

Few papers written in other languages had been accepted for many years, but not because they were written in French or German. In fact, few were being submitted in any other language than English to start with, because authors were aware that it would restrict a paper's impact in the scientific community, no matter how good the science. The Editors' report to the A&A Board in May 2002 shows this awareness:

> A few authors from France kept submitting papers in their national language. This causes two problems: first, papers in French have a limited readership in the international community and thus contribute to lowering the impact factor of A&A, and second they can only be sent to referees with a good knowledge of French, which
>
> 1. dramatically constrains the choice of referee,
> 2. does not always insure that the most competent referee is used for a given paper, and
> 3. creates a *de facto* inequality between A&A authors.

These problems led the Board of Directors to decide to modify the Instructions to Authors to make it mandatory for all submissions to be written in English. Until the first official hiring of a language editor in 2000, only the least understandable papers were sent to someone to have just the language corrected whenever needed, so that all the other papers must have been corrected with the help of either a friend of the author or the referee before final submission, or not at all.

2.1 English as a Lingua Franca

English is the global language in many domains, not just science. It is the *lingua franca* at meetings where participants come from several cultures and speak many languages, much like Latin was in the West in the first millennium.

In my opinion, the significant reason English is used is not that, for example, the British, US North Americans, or Australians speak it as a first language. If this were the case, then we should all be using Spanish or Chinese instead because they have more native speakers than does English.

The deciding factor is that English has become the main second language of all the other cultures in those areas where global communication counts, and it has become so even more since the fall of the Iron Curtain. As a language, English is not any better able to adapt to this role than any other language, at least not linguistically better: it just happened that way in this millennium. Wikipedia has several articles summarizing this role of English, if the reader is interested in exploring it further[1]. That hegemony as a global language may change in future millennia, who knows?

In line with this use of English, A&A is not just a vehicle for publishing scientific articles. It also reflects the growing community of astronomers and astrophysicists from over 60 countries for whom publications in English are the main medium of communication. It takes that role as a forum seriously by setting up ways for less advantaged scientific groups to publish their research, and the language editing service has always played a part in this effort. Between the reader survey and the editorials about policies, it also believes in the benefits of communicating its procedures, as the reader will discover from the various materials intended to help the author through the language editing process, found online and at the end of this article, and by our availability to answer questions about why certain suggestions are made for the English expression.

2.2 Editing for Clear and Correct English Expression

Besides the main reason cited above for using English, A&A soon began to concentrate on improving the overall clarity of its published articles by introducing language editing to make certain a good level of English is maintained. There are several reasons for this, beginning with the obvious desire to be read and understood by the most possible readers in the field. I list some others here.

- Since language is edited in almost all professional publications in all languages, A&A is only showing its professionalism, thereby providing the best possible forum for the research.

- Most of the Journal's authors are not native English speakers, so this move by the Journal is an attempt to keep authors from being unfairly restricted

---

[1] See http://en.wikipedia.org/wiki/English_language#English_as_a_global_language

by the need to publish in another language. It is by no means an attempt to add just another step in the already complicated one of publishing.

- The job of the language editors is to allow the science to appear as it is intended. If the language is awkward or garbled, the science may be interpreted as unclear, too.

The language editors have been hired to help the Journal's articles show more coherence in the language and form of all papers, and this remains the goal, even if there are a few differences in how each separate paper is treated. To achieve this, the language editors rely on grammar rules and the precepts of good style – the plain sort – they were all taught and trained in from the start. And to a certain degree, they must depend on a "feel for what sounds natural", even if not able to define a specific rule for the suggestion.

3   What is Edited: Seeking Quality

The language editors do not try to affect the science in any way. That is the job of the referee and scientific editor. Instead they look at the language and some details of presentation, such as the figure or table captions, paragraphing, and the abstracts. The goals of A&A English language editing are to correct

- grammar and vocabulary, including inconsistent spelling;

- ambiguous sentences;

- colloquial or familiar language;

- awkward expression, so it is easy to read with smooth, flowing sentences that are not too long or complex.

In other words, the language should be correct, consistent, unambiguous, and formal, and it should also carry the content, so that the article reads smoothly and clearly. The language should flow as the reader continues through the text, with nothing getting in the way of understanding what is being said, any more than will occur for those whose English is not yet a high level.

Review by a trained, but nonspecialist reader allows areas of text to be detected that are not very clear or that may be confusing to some readers, including to new scientists or those not in the author's own specialty. To this end, the style of a scientific article should allow the reader to pay full attention to the scientific content and not to how complicated, elegant, or even awkwardly it is written.

Another factor is the attempt to create a professional context for presenting the author's work, which is also taken into account in our suggestions. Besides the goal of a serious and formal style, they also aim for consistent spelling and word use, which is another measure of a publication's level. The second section of the *Guide to English editing at A&A* states this principle and goes into detail on the main inconsistency to correct: mixing British and US American spellings. We do not require one or the other at A&A, but expect it be the same throughout each single article. It may be a detail as far as clarity is concerned, but is one that is considered important in the publishing world.

3.1 Why Simple, as Well as Correct, English

We read the text to assess its grammar, syntax, and clarity. We intervene first to remove grammatical errors and inconsistencies in spelling or word choice, second to resolve any ambiguities of expression, and third to smooth out or simplify the style. Editing for a simple and correct English basically means that overly poetic language with its metaphors and elegant phrasing is discouraged, unless the writer has full control over it and a good reason to use it. Every reader should be able to follow the ideas without difficulty. In this case, "simple" does not mean "simplistic" or "simplified", but rather:

- use clear and plain rather than elaborate language;

- use the main dictionary meanings of words, not their literary or extended senses;

- avoid metaphor, jargon, or long convoluted sentences.

This is the opposite of what it may seem when your words are being rearranged and alternative phrasing is suggested. This is not the style of, say, Shakespeare, in his poems and plays. Since he was writing before any grammar or spelling books had been written for English, unlike Latin or Greek, the language editors would now be forced to correct his originals much more than any science article seen nowadays, but they would hardly try to reproduce his specialty, the exquisite poetry and rhetoric, as in

> Thus Conscience does make Cowards of us all,
> And thus the Native hew of Resolution
> Is sicklied o're, with the pale cast of Thought,
> And enterprises of great pith and moment,
> With this regard their Currants turne away,
> And loose the name of Action[2].

But A&A would suggest he change to the literal expressions, not the poetic figures of speech, not to mention all the spelling, word choice, and capitalization problems this contains (*e.g.*, hew, Currants, turne, loose), which no longer show up in modern editions.

Even then, the editors make certain that it will be understood by all readers, regardless of linguistic background. Poetic language is by nature ambiguous with its multiple layers of meaning. This is the policy for all publications of scientific journals in the United States and the United Kingdom, if not elsewhere.

---

[2] *Hamlet* Act 3 scene 1, Folio I (1623).

In the very useful *Style Guide of the American Chemical Society*, which is run in conjunction with Oxford University Press[3], I find two quotations on this question:

> Long words and complicated sentences are not essential features of good scientific writing, although they are often thought to be so. The best writing in science, as elsewhere, is simple, clear, precise, and vigorous. Decide what you want to say and say it as simply, informatively, and directly as possible. (M. O'Connor, *Writing Successfully in Science*),

and

> In scientific writing, there is no room for and no need for ornamentation. The flowery literary embellishments, the metaphors, the similes, and the idiomatic expressions are very likely to cause confusion and should seldom be used in writing research papers. Science is simply too important to be communicated in anything other than words of certain meaning.

The second citation is from Robert Day in *Scientific English: A Guide for Scientists and Other Professionals*, and it goes on to say that "The meaning should be clear not only to peers of the author, but also to students just embarking on their careers, to scientists reading outside their narrow discipline, and especially to those readers (the majority of readers today) whose native language is other than English".

In this spirit, A&A also asks that the jargon and specialist phrases used among collaborators be avoided except when there is no other choice, because it may not be as familiar or clear to the other scientists reading the article; likewise, they are never as precise as the standard form, unless it is the only technical way it is expressed. If the language editors question the use of words that you feel are the most suitable, they are not challenging the science, but have instead seen a problem in the phrasing or effectiveness of its presentation. If a standard set of words exists in English for the same idea, then please use it instead of the specialist phrasing, even if it is present in most of your colleagues' papers, or try to vary between the two.

Likewise, there are some wordy expressions that show up in scientific and other specialized writing, so that untrained writers seem to feel they are more precise or more scientific than the standard form. Indeed, they are found in native English-speakers' papers, so technically correct, but it is perceived as bad style and unprofessional. When there are too many in the same section, it is either confusing or awkward.

In English, the main principle is that the more unnecessary words used, the less clarity in the message. Examples of some phrases include

---

[3] http://www.oup.com/us/samplechapters/0841234620/?view=usa

be in agreement with, of the order of (or *on, in,* depending on the dialect), *small number of, despite the fact that, show evidence of the presence of, owing to the fact that* (or any 'the fact that'), *are known to be, at the present time, of great importance.*

Each of these should be replaced by their direct and shorter synonym as follows in the same order:

agree with, about (*roughly,* etc.), *few, although, proves the presence of* (or *evidence of* without *presence of*), *because, are, now, important.*

As you can see, even with the parentheses (similar character counts in both), the second list is much shorter than the first. This list was adapted from the much longer one on the Oxford and ACS page mentioned above. There are others we see in A&A submissions that are listed in our resources:

- the appendix to the October 2008 editorial on language editing (see the News page of the A&A website)

- "Some Frequent Corrections,"[4] or

- Sect. 6.5 of the *Guide to English editing* on the site.

That conciseness is an ideal in all or most English writing is supported by several experts and is one of the main things taught in composition classes in all English-speaking environments, so aiming toward that goal as much as possible is another consideration in the language editing, although clarity is still the main one. The most famous statement of this principle comes from William Strunk, Jr. in *The Elements of Style*:

A sentence should contain no unnecessary words, a paragraph no unnecessary sentences, for the same reason that a drawing should have no unnecessary lines and a machine no unnecessary parts. This requires not that the writer make all his sentences short, or that he avoid all detail and treat his subjects only in outline, but that every word tell.

It has been shown over and over that using more words than needed leads to confusion so that the point in a sentence, paragraph, or article is lost. The *"Guide to English editing"* and *The Elements of Style* offer several other suggestions for writing more concisely and clearly, including examples of how to make your sentences more active or to express parallel ideas in the same structure.

3.2 Why Formal English?

Asking for formal English stems from its being standard across all the national dialects of English, which is not the case for the informal versions of English.

---

[4] At http://www.aanda.org/language-editing

All standard formal versions of language aim at communicating clearly across differences, while familiar language like slang is meant to be specific to a single group or place, *i.e.*, to exclude others from the message. Informal or colloquial English creates a slightly different problem for both writers and readers than elaborate language, because it

- cannot be easily understood in all regions. There are too many variations that do not carry the same meaning between regions and dialects, even in the same language. This is not the case with the formal standard written language.

- is hard to use correctly or precisely. This is true for native speakers, so even more so for non-natives, and informal language changes too quickly for it to be used in archival documents, like a scientific journal should aim to be.

Once again, Wikipedia's writers summarize the situation very well:

> Formal written English is a version of the language that is almost universally agreed upon by educated English speakers around the world. It takes virtually the same form no matter what the local spoken dialect is. In spoken English, there are a vast number of differences between dialects, accents, and varieties of slang. In contrast, local variations in the formal written version of the language are quite limited... The differences in formal writing that occur in the various parts of the English-speaking world are so slight that many dozens of pages of formal English can be read without the reader coming across any clues as to the origin of the writer, far less any difficulties of comprehension[5].

4   The Role of Language Editing in the Editorial Process

Language editing is an integral part of the editorial process at A&A. When a paper is first submitted, the scientific editor may choose to send the paper back to the author requesting that the level of language be improved before sending the paper to an external referee. This is usually because the editor has determined that the paper contains enough scientific content to warrant review, but also that the level of English hinders appropriate evaluation or that the paper will require undue language revision to reach the level sought at A&A.

This step acknowledges the difficulty of having to express oneself clearly in written English, regardless of one's first language, and the disadvantage relative to native English speakers of being required to publish in English. Having a manuscript clarified by English-language colleagues before submission can, therefore, speed up the publication process.

Before acceptance, the referee may give advice for improvements in the language along with the scientific presentation, or else recommend it for language

---

[5]See http://en.wikipedia.org/wiki/English_language

editing at the Journal. Editing language is not the main function of the referee, but some do it as they are commenting on the article. Before language editing was made official, the referee played a larger role in this, but not all referees feel qualified to correct English, even when their level is higher than the authors' in English. Instead, they are now only asked to make recommendations for whether revision of language is needed.

When the paper is being accepted, the scientific editor may make a recommendation that a paper be edited for the English and this is considered in the final judgment. A&A now has a single language editor who looks at every single paper for the quality of the language in the article. Besides scanning them to see whether the language needs revision, this language editor also looks closely at all abstracts for correctness, clarity, and conformity to the goal and style of an abstract. At this point the paper is then (i) sent on for language editing, if needed; (ii) sent on directly for publication; (iii) after the first page is corrected, sent back to the author for a final pass (sometimes with some suggested changes later in the article). (iv) In cases where there is a native English-speaking fellow author, she or he might be asked to look through it one last time again for minor corrections.

4.1 How to React to Suggestions

Corrected versions are returned to the authors for confirmation and corroboration. The author then should incorporate those changes into the LaTeX file of the paper and resubmit for publication. At this point, there are 3 choices.

- You agree with all the suggestions and resubmit to the office in Paris following the instructions in the email from the language editor. If there are no other problems, it will be sent on to the publisher directly. At this point it is copyedited by the publisher and the proofs sent to you for confirmation.

- You remain uncertain about some of the suggested changes and would like further explanations. Write an email with your questions to be sent to the language editor, who will answer you as soon as possible.

- You do not agree with a proposed change, because it changes your meaning, especially the scientific import. If a change was suggested, it means that the original has some problem in the expression, and the suggestion was the most logical with the words being used in the original.

If there are several such issues, it is helpful to have the preferred changes marked in bold face in a referee format version of the text uploaded on the A&A MMS website. The preferred corrections are almost always accepted provided no additional language errors are introduced. In matters of scientific content, the author's preference is always respected. It is also likely that the language editor will take the time to explain the suggested changes you are questioning, if your new version is not correct English. Some of the explanations can be found in the *"Guide to English editing"*, of course, which was originally designed to be used at this stage.

5   Other Aspects to Look at

When reading a paper for the English expression, the language editors also take a look at the abstract, figure and table captions, equation punctuation, and the reference list to make certain they conform to the principles of the *Instructions to Authors*.

Typical problems that are dealt with in abstracts are

1. an incomplete abstract or one that is not in the expected order (the headings), or one where, for instance, the context is too long but *Aims*, which is obligatory, is too short.

2. an incomplete sentence when the headings are used, because the author follows *Aims* or *Method* as if it were a question, *i.e.*, the noun phrase that describes the aims or method. We expect each sentence to be a full sentence, such that the whole abstract would read coherently if the headings were removed.

3. references, including self-reference, in the abstract, which should be self-contained, that is, should only refer to the paper that follows, not to the background information expected in the introduction.

There are exceptions to this last rule (a direct dependence on another work, as when commenting on a specific result published by another group rather than simply working from their results, as all science does), but it should be avoided whenever possible. If another paper must be cited, then the full reference should be included, since abstracts are often published without their papers and reference lists. The editors will query every single reference in an abstract and ask the author to check with their scientific editor or the Chief Editor if there is a question of appropriate reference. If the present paper is a follow-up to an earlier paper in a series, the "Paper I in the series" is enough to use in the abstract, without the full reference.

The main problem found in the table captions is when authors try to include as much as possible about it: all the references, legend details, full description of column headings and of results. If an author is tempted to say "Note how the..." then it is discussing results that belong in the running text not the caption, which should refer exclusively to the legend symbols in the figure. There is a difference between table and figure captions because a table has a title and notes, while figures have short captions that explain the graphical information. There should be a minimum of repetition between text and caption, the exception being the brief description of the subject of the table or figure (in title form without many details) followed by the minimum of what is needed to follow the figure or table in a legend form. The Instructions to Authors give more specific instructions, but the only exception to these long captions are for figures that are published separately online, so that the full descriptions of columns and legends and references are needed, because the text is not directly available, such as it is when it is embedded within a full text.

Even though the equations are part of the scientific editing, it follows the syntax of the sentence that it must fit into, what is called embedding, so is also part of the language. They may never stand alone, but instead continue the preceding sentence as an example following a colon (":") or directly with no punctuation at all before it begins, even when separated. The sentence can continue after the equation if needed, but then no colon or semi-colon (";") should be used more than once, as in a normal sentence without any equations. Likewise, when the sentence ends with the equation, then a period (full stop) follows it directly. This is the house style, so it will also be checked by the copy editors at the publishers.

The reference list is glanced through for potential problems, although a more thorough job is done at the publishers. If there are arXiv references without indicating what the paper's status is with the journal where it will be published, then it will be queried and a reminder goes to the author to make certain that this is included in the final version. There are few cases where A&A accepts work that has not been refereed, so these cases will need to be gone through with an editor.

6   Recommended Resources when Writing for A&A

The following resources are recommended for answering questions about suggestions for changes and why they are made. The language editors have been working together to provide you with most of the guidelines we use when correcting your English expression. The *Guide to English editing at A&A* is for your use at any moment. It does not cover every single rule of English usage (*e.g.*, use of articles for languages that have none or few, because it cannot, but it discusses those that are found most across all the authors' linguistic backgrounds.

1. Author's guide on the A&A website, starting with the Instructions to Authors. This covers everything except language use. Every author should refer to it regularly while writing and before submitting.

2. The other pages for authors, including the "Language editing" pages and its attachments. These include the full guide to English editing, and the attached file includes a PDF file that is easy to download for future use: *Guide to English editing at A&A*. You can also go to the last section for how to correct some phrases we correct a lot in A&A articles and what we are likely to suggest as a better way to say it. The explanations for many of them are in the *Guide*, but not all, since many are vocabulary problems, such as which preposition goes with which noun or verb.

3. Your own bilingual dictionary but complemented by a monolingual dictionary, if possible. You can always use the online versions of (a) the Cambridge dictionaries, where they will also tell you which are the British and which

the American spellings[6], or (b) Webster's online with input from Princeton and INSEAD (*Institut Européen d'Administration des Affaires*) professors[7].

Guidebooks from other scientific disciplines and laboratories that follow the same principles as A&A so you can use them for most questions, even if their examples come from their disciplines. These include

- CHEMISTRY: The ACS Style Guide: A Manual for Authors and Editors, Second Edition[8], edited by Janet S. Dodd. Its sections are: Getting Started, Writing Style and Word Usage, Components of a Paper, Types of Presentations, Advice from the Authorities.

- BIOLOGY, the biology department at Columbia University "WRITING A SCIENTIFIC RESEARCH ARTICLE"[9]. Its main sections are found under "Format for the paper" and "Edit your paper!!" Both of these repeat in detail what language editors are looking for when editing your papers, so you can see there that we are not being any more rigorous than others are.

- NASA – Langley Research Laboratories guide for authors presenting reports by Mary McCaskill[10]. This is the third NASA URL I found for this resource, but sadly it was not working as of the latest version of this chapter. Try to find it by a search engine using the title and author; if not, consider the following list of English for science resources to complement those here: `http://webster.commnet.edu/writing/writing.htm`. This resource was built by Capitol Community College, which lists its own very useful interactive style and grammar guide first.

- The latest edition of *The Chicago Manual of Style*, Chicago University Press. This extensive manual is used by most scientific communities in the United States, including psychology. It may be too technical and detailed, when most of what you need is on the A&A site. The book itself is bulky, but it is possible to subscribe to it online for $ 30 (US) a year to access all its sections (including the style guide) and even to ask specific questions of its staff of editors.

Other resources online for scientific writing and writing in general:

- Madison Wisconsin Writing Lab for Scientific Reports[11], including suggestions for each of the 6 parts of a paper (or scientific report).

- A list of common errors and advice by a professor of English at Washington State University, Paul Brians. It tells you it is meant for native English

---

[6]`http://dictionary.cambridge.org/`
[7]`http://www.websters-online-dictionary.org/`
[8]`http://www.oup.com/us/samplechapters/0841234620/?view=usa`
[9]`http://www.columbia.edu/cu/biology/ug/research/paper.html`
[10]`http://www.sti.nasa.gov/publish/sp7084.pdf.`
[11]`http://www.wisc.edu/writing/Handbook/ScienceReport.html`

speakers[12], but there is much that can help everyone. He also gives a list of sites specifically for second-language English writers.

- Purdue University Online Writing Laboratory (OWL) pages for second language students of English[13].

- English-at-home.com (go to "grammar").

7 Concluding Statement

The search for coherence and quality includes approaching the style that is considered to be best in the language being used, and in English this is best expressed by Strunk & White's *Elements of Style*. If science writing considered that using the jargoned clichés from its informal style were better than plain and standard English, then the style guides for the various scientific disciplines would not refer back to this reference work over and over, as done here.

At times while working through a manuscript I empathize with authors who had first learned a correct and clear English to express their findings. As time went on, however, they picked up these phrases in the literature and at conferences, learned what they meant, and supposed that these were more appropriate for some reason. Since scientists use them, they must be more precise; or perhaps since an English speaker uses them, they must be more correct, a writer might think. Many are correct, admittedly, since some or most are found in a dictionary, which is appropriate, but most of them are either bad style or hard to use correctly.

The work that has already been put into the papers is considerable before it gets to the language editors, and the English is usually clear enough to make suggestions to make it more effective, as well as more correct. If this work on the English had not been done by authors before submission, little could be suggested beyond basic grammar and vocabulary corrections alone. Going beyond this first level to suggest changes for clarity, consistency, and a smooth style is intended to move in the direction of improving the impact of the research published in A&A.

The response to language editing from the majority of the authors who responded to the Author Survey of 2007 supported the Journal's efforts, just as most authors respond very positively to these efforts by either incorporating the suggestions, by explaining and communicating constructively when not understanding or agreeing with some of the suggestions, or even by sending information on technical expressions that helps the language editors in their task. Communication is, after all, the goal from the time the observations and calculations are completed to when this work and its results need to be expressed and disseminated to colleagues at home and abroad.

This work is based on all the collaborative efforts of the language editors and staff at A&A through the years.

---

[12]http://www.wsu.edu/~brians/errors/
[13]http://owl.english.purdue.edu/owl/resource/678/01/

THE ROLE OF THE PUBLISHER

Agnès Henri[1]

Abstract. This paper describes the various facets of the publishing activity. The task of the publisher is to make every effort to promote the work of researchers who have submitted their article for publication. This requires the set up and maintenance of an effective management interface, timely production, delivery of quality items, a high-performance web interface, sustainable data archiving, indexing in large databases, and round-the-clock maintenance of data access. It is a daily work that envolves a dedicated professionals team for the timely production of *Astronomy & Astrophysics*.

1   Introduction

Managing a journal, and especially a major astronomy journal like *Astronomy & Astrophysics*, requires time, energy, and professionalism. Most people think that typesetting is the prominent part of the publishing enterprise; because the authors use LaTeX macros, the publisher no longer seems really necessary. Although it is true that, compared to twenty years ago, the publisher is less involved with the layout, a lot more time – hence expenditure – is now required for the editorial management, the online publication, the archiving and indexing of the data, etc. EDP Sciences (hereafter EDPS) has been publishing physics journals and series of books since 1920, and thus has acquired sound experience in the production and publication of scientific journals.

EDPS is a subsidiary of learned societies and, as such, its prime aim is to participate in the dissemination of scientific information, rather than the mere pursuit of profit. For the research community as a whole, community-owned journals represent the best vehicle for distributing the scientific findings, are the optimal media for promoting the scientist's work, and last but not least, act as significant catalysts for the individual research careers. The publisher is the partner of the scientific community, and the partner's role is to work in collaboration with the editorial team of scientists on the following main points:

1. improve the papers during the refereeing process with the technical services of the publisher,

---

[1] EDP Sciences, 17 avenue du Hoggar, BP. 112, 91944 Les Ulis Cedex A, France

2. be a reliable source of information (only validated versions are published), and

3. promote the work and career of researchers.

Our mission is therefore to provide the A&A scientific community with all necessary tools for enhancing the visibility of their work at the lowest possible cost for the subscriber.

The following sections describe the different parts of a publisher's job: from the management tools to online publication, through production, marketing and web development.

2 Management Tools: From Submission to Acceptance

To produce a journal such as A&A, which receives around 2500 submissions per year, the editorial office needs a high-performance tool for managing all documents. Indeed, the editorial office of a journal is the very first contact that scientists have when they submit their manuscript. In this respect, it must guarantee efficient service and satisfaction for authors. An efficient software tool is needed for submitting manuscripts and following the peer-review and handling of articles by each scientific editor.   The A&A editorial office uses the *Manuscript Management System* (MMS), a manuscript management software package developed by EDPS, which was specially designed to help editors, authors, and editorial staff manage in real time the manuscripts under evaluation. The maintenance and development costs for this tool involves two positions (software engineer and product manager) at EDPS.

3 Production: From Acceptance to Publication

A high-quality scientific journal must feature efficient, high-quality production and a fast publication time.   The production involves different steps that are summarized in Figure 1. The production process is more and more managed like an industrial process, with the aim of reducing delays and costs.

3.1 The First Step: Copy Editing

After acceptance of a manuscript, the authors send the files of the manuscript to a dedicated e-mail address or via the MMS. A team of production editors takes care of the manuscripts with regard to

- check of the spelling, typographic mistakes, etc.;

- preparation of the references for use by CrossRef[1] in the online version;

- consistency check of graphics; and

---

[1] CrossRef's mandate is to be the citation-linking backbone of all scholarly information in electronic form, see http://www.crossref.org/

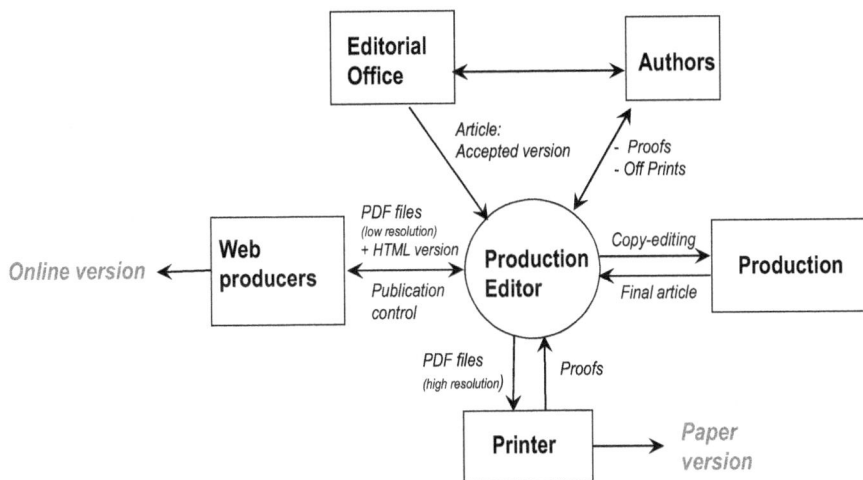

Fig. 1. Production workflow from acceptance to publication.

- content consistency: make clear distinctions between physical variables, mathematical symbols, units of measurements, abbreviations, chemical formulas, etc.

We mentioned earlier that compared to two decades ago, authors now do a lot of typesetting labor when preparing their manuscript by using, for example, the LATEX macros provided by the publisher. Nevertheless, they are not fully aware that, quite often, small errors remain: authors are not expert typesetters and, because they concentrate more on the content than on the form, they make mistakes in the placement of figures, they incorrectly quote page and volume numbers in the bibliography, etc. Figure 2 (which should be viewed in color) shows an example of a copy-edited bibliography page in A&A. It clearly demonstrates that numerous corrections have to be made in order that links towards other journals work.

3.2 The Second Step: Prepress

Once the manuscript has been proofread, the EDPS typesetter brings in the corrections required by the production editor and applies the layout with the A&A LATEX macro. For the figures, the typesetter verifies consistency of graphics (*e.g.*, by cleaning the figures, correcting the textual components in graphs, adjusting thickness of lines, checking the colors – *i.e.*, RGB or $CMYK$ depending whether the figure is for printing or for the online version, incorporating all fonts needed to avoid any problems during the final printing process, fixing the resolution, etc.). Once this process is finished, the manuscript PDF file is returned to the production editors. They then check that everything is correct and send a proof to the author for final correction and approval. Once the corrections by the author have

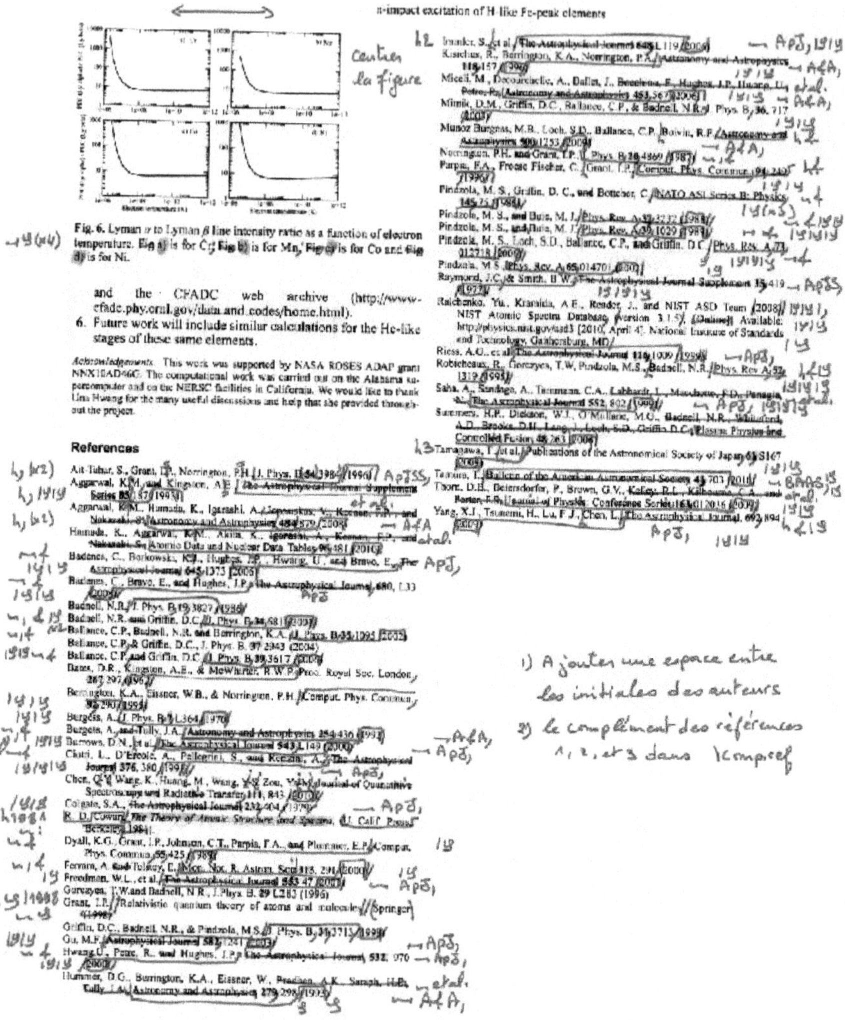

Fig. 2. An example of copy-editing of the bibliographic references in A&A.

been applied and after a final check of the paper's contents by the editorial office, the production editor inserts a *Digital Object Identifier* (DOI) number[2], and the article and volume number. The article is then ready for publication.

---

[2]A DOI number is attributed to each paper (http://dx.doi.org), ensuring the longevity of the article's URL. Together with CrossRef, the DOI identifier also allows for including a direct link to the publisher's website of the referenced articles in the online version of the bibliographic references.

3.3 The Third Step: Publication of the Online Version

The files are then processed for the web version: preparation of metadata, full-text and references in XML (the XML format allows for the creation of the HTML full-text web version, and is currently the format needed for data sustainability). In the newly introduced system where articles rather than pages are numbered, the journal volume is built up in a progressive manner and is brought to a close when an adequate number of pages have been published.

3.4 The Fourth Step: The Paper Version – Printing and Dispatching

When enough articles and pages have been produced to create a volume, the production editor prepares it (by organizing contents, cover, etc.), and sends it by ftp to the printing house. Production editors are in permanent contact with the print manager to solve any technical problems that may arise, and each stage of the printing process is checked.

4 Web Platform

Maximizing the distribution yields more visibility to articles and improves the impact factor of the journal, thereby promoting the results presented in the articles, their authors, and the institutions that employ them. What is therefore required is a state-of-the-art, efficient web platform offering numerous services and advanced features for distributing the online edition of the journal. The A&A platform includes several tools, as follows.

CrossRef indexing and hyperlinks in the bibliographic references. There are approximately 2550 participants in the CrossRef system (publishers, research companies and institutions, libraries), representing more than 20 000 journals. The system enables the creation of millions of links between the different journals. EDPS also provides direct links to referenced articles in other databases such as NASA/ADS and MathSciNet. Readers appreciate these features, because they allow them to easily navigate between different articles in a single domain issued by different publishers.

Online publication of the HTML and PDF versions, electronic-only material, and archiving. As we described above, the publisher should provide a full-text HTML version of each paper, and also allow for electronic-only material designed to provide supplementary information that is either too voluminous for printing, or designed specifically for the Web, such as large tables, appendices, programs, images and movies, etc. Another mission of the publisher is to archive the contents and to ensure permanent access to the journal, by

- hosting the data for an indefinite length of time;
- providing XML archives and PDF files;

- retro-converting data from time to time in order to maintain the accessibility to the contents;

- delivering abstracts to databases such as ADS-NASA, ISI, and Scopus;

- linking to data available at the CDS, and

- providing a platform that gives rapid access, even when traffic levels are high, with site maintenance ensured 24 hours out of 24, 365 days a year.

Specialized services – reader information. The publisher's platform should offer a wide range of services for the reader:

- Email alert and RSS feeds.

- "Citing article": this returns a list of articles that cite the selected article. The citing articles come from the EDPS Electronic Journal database, as well as from other publishers participating in the CrossRef Forward Linking System.

- "Articles from the same author": this provides a list of papers written by the same author in the EDPS database.

- "Related articles": this furnishes a hyperlink to articles with a common author or common keywords in the EDPS database.

- A website fully indexed by Google, and participating in Google Scholar[3].

- Database collaboration, *i.e.*,

  - CDS: EDPS, in collaboration with CDS, improves the author-supplied links pointing to astronomical objects in the CDS databases and virtual observatory tools.

  - ADS-NASA: EDPS supplies the complete archives of A&A Supplement, returns XML code conforming to the ADS technical specifications for each new issue of the journal.

- Dexter application: ADS has developed the applet Dexter, which has been adapted by EDPS to work with A&A articles. This software provides the capability of exporting data from the figures.

- Search engines:

  - CrossRef Search: this pilot system, put in place by the CrossRef consortium, is a free search engine that facilitates navigation in scientific journals by specifically indexing scientific articles. The search engine uses Google technology and DOIs.

---

[3]Google Scholar is the academic version of Google.

    – Intuition search engine: this advanced search engine provides multi-criteria searching: author, title, keywords, addresses, publication date and complete text. Searches can be made in a particular journal, in a particular scientific domain (physics journals, for example), or in all the journals published by EDPS.

- OAI Server (Open Access Initiative): EDPS has put in place an Open Access Initiative depot, which will make available articles metadata in a standard format (DublinCore) at no charge, thus enabling the interoperability of different OAI archives. Furthermore, the Publisher must permanently analyze and evaluate new emerging technologies so as to offer more and improved services to the journals and their readers.

The maintenance of the web platform including web hosting, software management, and software development keeps a team of several people busy at EDPS.

5 Communication and Marketing

The marketing and communication department's aim is to maximize the journal's visibility to researchers, authors and readers. The more people become aware of what happens in the journal (publication of high-quality papers, special issues, etc.), the more people will read and cite its articles, thus bringing to the authors the visibility they need. A team of marketing assistants must

- print promotional material (a leaflet for Astronomy & Astrophysics) and distribute it as widely as possible,

- carry out promotional items that will disseminate additional support materials for authors and readers,

- send information and announcements to readers (on a new section, a high-lighted paper, etc.),

- represent the journal at international scientific meetings (AAS meetings, IAU General Assemblies, SWYA school, etc.),

- promote selected articles: some outstanding articles merit being brought to the forefront by the editors. For these articles, EDPS works in partnership with the editorial office, to

    – make them freely available,

    – highlight them by a specific category ("Press releases" or "Highlighted Papers") that is directly accessible from the homepage of the journal,

    – add an accompanying text to explain the originality of the article.

6 Conclusion

The publishing business has changed significantly in recent years. Expertise has shifted to activities that are much wider than layout and dissemination of the printed edition of an article. Advanced technological tools are required (article management tools, an up-to-date publishing platform) with specialized technical personnel. A dedicated team of several professionals works at EDPS to give A&A the highest impact.

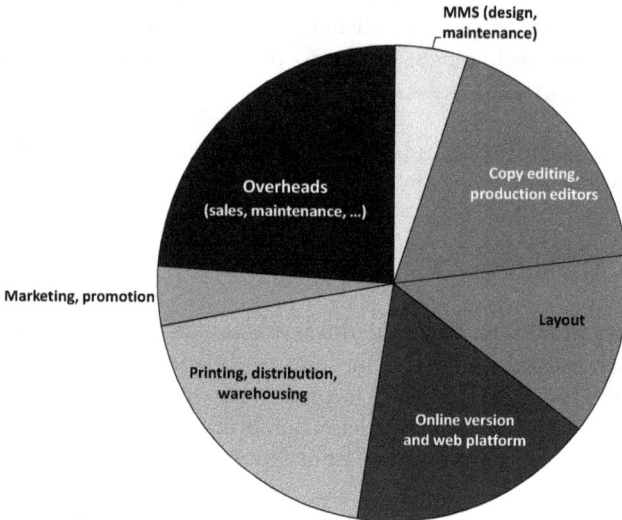

Fig. 3. Sample distribution of publisher cost factors (*i.e.*, non-editorial publishing expenditures).

The list below summarizes the cost factors on the publisher's end, and Figure 3 illustrates the relative shares of each item in the total operational cost.

1. development and maintenance of the Manuscript Management System;

2. copy editing as a consistency check of all aspects of the manuscript (figures, references, etc.);

3. prepress work (layout);

4. publication of the online edition (PDF, HTML, online material, . . .), electronic only figures and tables, etc.;

5. printing and dispatching;

6. archival services, and

7. specialized services (dedicated web site with many functionalities, marketing, etc.).

I thank Chris Sterken for proofreading the paper.

ASTRONOMY LIBRARIES – YOUR GATEWAY TO INFORMATION

Uta Grothkopf[1]

Abstract. This chapter reviews services offered by astronomy libraries for assisting astronomers in their research. Special attention is given to the two most important information tools in astronomy: the NASA Astrophysics Data System (ADS) and the arXiv e-print server. Some less known features will be presented to lead to more efficient use of these tools. The core astronomy journals are explained, along with the open access concept and how it interacts with existing journals. The chapter also provides an introduction to bibliometric studies and shows why publication and citation statistics are important for both researchers and observatories. Finally, examples of cooperation between librarians and astronomers are presented in the context of working groups and conferences.

1 Introduction

The world of libraries is changing fast. The times when they were simply repositories of printed documents are long gone, because preserving the astronomy legacy and being curators of historic documents is important, but is not the only challenge librarians meet today. In the internet era, librarians deliver services directly to the scienstist's desktop, either by providing access to scientific journals and databases through assistance in using specific information tools or simply by answering reference questions via email or messaging systems. Astronomy libraries are a good example of this changing paradigm, because the vast majority of information resources are available electronically, and most astronomers are used to, and actually prefer, online resources over printed formats.

To use information efficiently, students and researchers need to acquire some basic skills. These can be divided into three parts.

- *Information literacy:* We have a surplus of information rather than a shortage. Scientists have to be able to locate the texts that are important for

---

[1] European Southern Observatory, Karl-Schwarzschild-Str. 2, 85748 Garching, Germany
e-mail: `esolib@eso.org`

their research, otherwise they may waste time with information they do not actually need. Effective use of finding aids like library catalogs and reference works is called information literacy.

- *Computer literacy:* A certain understanding of the hardware, software, and network technology is needed to access information tools from the office, from home, or from abroad. This knowledge is called computer literacy.

- *Critical thinking skills:* Regardless of the medium (print or electronic, local or networked), no search result should be accepted without questioning its validity, correctness and completeness. Such critical thinking skills include common sense, as well as techniques for fully understanding a search result and putting it into context.

The combination of information literacy, computer literacy, and critical thinking skills is defined as information fluency, as explained in Viele (2008). Librarians will help develop these techniques and thus make the most of the time spent on literature and information searches.

2   What Librarians Can Do for You

In today's interconnected world, a large fraction of the scholarly resources in astronomy are available online. Latest preprints are retrieved from the arXiv (astro-ph) e-print server, journal articles can be obtained via the NASA Astrophysics Data System (ADS) abstract service, and communication with colleagues of course mostly takes place by email. When asked what information services they use in their daily work, scientists often do not immediately think of their library. Yet, most astronomers do use libraries, even though they may not always realize it.

Start thinking of libraries not only as collections of printed books and journals, but also as gateways to electronic publications. Every time you download a recent journal article from the ADS, most probably you are allowed to do so because your library has paid the necessary subscription fee, negotiated access rights, and does the troubleshooting every time access is interrupted.

Librarians also compile web pages with link collections that provide access to electronic books, reference works, and databases. They help scientists discover new information retrieval systems and assist them in their research. To do this, librarians form professional networks to overcome the limited local resources and make use of joint collections, knowledge, and ideas provided by astronomy librarians worldwide.

In many organizations, librarians are in charge of institutional repositories. They hold introductory sessions to explain how to deposit manuscripts in the repository, make sure all necessary metadata get submitted, and are often experts in copyright and Open Access (OA). They also maintain databases of all papers published by the organization's scientists or, in the case of observatories, of those papers that use observational data from specific facilities. From these databases, statistics are derived and made available to management, governing bodies, and the public.

The library mission traditionally has been to fulfill the information needs of their users by selecting, collecting, preserving, and providing access to relevant resources. While many changes have occurred regarding the tools used in library management, this mission is still valid. In the digital age, personalized service, tailored to the needs of the library users, distinguishes libraries from software tools, see Grothkopf (2003).

Increasingly, research libraries act as meeting places. They provide space for both informal and formal meetings, ranging from the "science coffee" that takes place every morning to regular discussion groups and formally organized talks and presentations. Instead of the stereotypical quiet reading room, libraries often host collaborating scientists who appreciate the relaxed and inspiring atmosphere in the library.

Similar to the virtual "Web 2.0" concept, the traditional library has been replaced, or rather supplemented, by the "Library 2.0" that offers web pages with library blogs, RSS feeds, and interactive tools for contacting the librarians. Of course, especially in small, specialized libraries like those that often can be found in observatories, the most interactive way of consulting librarians is still to simply meet and talk to them. They are happy to provide research assistance, help you with finding quality information resources and suggest all kinds of tips and tricks you may not have heard of.

3   Two Astronomy Gorillas

Without doubt the majority of astronomers cover almost all their needs for scholarly information using only two resources: the NASA Astrophysics Data System (ADS) and the arXiv.org e-print server. ADS and arXiv clearly are the "gorillas" among their peers as the largest, most important, most visible resources. And they are both free of charge to the individual astronomer, hence available to everybody, no matter whether affiliated with an institution or not, or regardless how wealthy that institution is.

In the following, we briefly introduce both services and point out some features for using them efficiently.

3.1   NASA Astrophysics Data System (ADS)

The NASA ADS abstract service is operated by the Smithsonian Astrophysical Observatory (SAO) under a NASA grant. The main site is at `http://adsabs.harvard.edu`[1], with several mirror sites located around the world. It is the largest digital library in astronomy, providing links to more than 7 million records in astronomy, astrophysics, and physics, and it is used almost daily by the majority of

---

[1] ADS main search screen: `http://adsabs.harvard.edu/abstract_service.html`

astronomers. Access to subscription-based publications (including the two or three most recent years of the core astronomy journals) is granted only if the user or his/her institution subscribes to that publication. For the years 1975 through approximately 1995, the abstracts database contains data from a variety of sources; since 1995, most abstracts have been provided directly by publishers or journal editors. These references should be 100% complete. With less completeness depending on the subject area, but still at a high rate, ADS also provides information about older literature and, in many cases, offers links to scanned articles back to volume 1 of the most important astronomy journals. In addition to published literature, ADS provides information about preprints posted on arXiv.org. Preprints can be retrieved via the ADS the same day they become available on arXiv, so that there is no delay in finding the most recent postings.

ADS receives information about publications from a variety of sources, most notably directly from publishers on or even before the publication date, as well as from the arXiv preprint database. ADS users can query the database for immediate results or subscribe to the myADS service to receive regular notifications about new publications that correspond to search criteria set by the user.

3.1.1   Some Search Tips

The ADS main search screen is mostly self-explanatory. However, there are some features users do not necessarily know about. Some of them are explained below, and they have been available in the ADS since January 2009. Since the ADS is continuously being developed and improved, further capabilities can be expected in future.

First author searches: To limit searches to first authors alone, use ^ before the author's name. The resulting list will only include papers with that person as first author.

Filters: By scrolling down on the main search screen, you will reach the Filters section with a variety of options. For instance, to retrieve only publications from refereed journals, select "All refereed articles" in the first Filters section ("Select References From"). By using the "Select/deselect publications" option, users can enter journal codes of publications that are to be included in (or excluded from) the results. For a list of journal abbreviations, see the ADS Frequently Asked Questions page[2].

Citations: A very useful addition to the bibliographic record is the information about citations. To see the number of citing papers, go to the "Sorting" section towards the end of the screen and select "Sort by citation count" (or "Sort by normalized citation count" to normalize by the number of authors). Then enter your search. Alternatively, you can first enter your search and then use the "Sort

---

[2]http://adsabs.harvard.edu/abs_doc/journals1.html

options" pull-down menu in the upper right corner to sort by citations. On the results screen, you will see the number of citations above the title of each entry. It will be displayed with three decimal places (*e.g.*, 32.000) which are only important for the normalized citation count when citation fractions below 1 can occur.

For any given result set, you can get the list of citations for all or selected articles of the result list. The citation lists can include or exclude self-cites, and they can be limited to citing papers in refereed journals. In order to use these features, first enter a search on the main search screen. Make sure the whole set of results is displayed; if necessary, the number of items to be returned can be increased (default: 200). On the results page, select the records for which you want to see the citing articles, or scroll down below the last record and choose "Select All Records". Then scroll down further until you find the option "Get citation lists for selected articles" or "Get refereed citation lists for selected articles". To exclude self-citations, check this option.

A detailed description of the creation and use of citations in the ADS has been given in Accomazzi *et al.* (2007).

Export search results: Next to the Sorting section on the main ADS page, you will find the Format area, which governs the output format of search results. Instead of the default short list format, a variety of formats can be selected that allow easy importing into bibliographic tools like EndNote and ProCite. To add records to a Zotero library (see also Sect. 9 on organizing your own library), users can simply click on the paper icon that shows up in the URL field of the browser window. ADS records can also be displayed in a format ready for cutting and pasting into TeX manuscripts, *e.g.*, BIBTEX or AASTeX. Reference lists already formatted for specific journals can be generated, for instance for *Icarus, MNRAS,* and *Solar Physics.*

Historical literature: In addition to current publications, the ADS also provides access to historical literature. Together with the John G. Wolbach Library at the Harvard-Smithsonian Center for Astrophysics, the ADS digitizes microfilms from selected historical publications in astronomy. Although this compilation still lacks pagination so that one cannot go directly to a specific page, it still is very useful, in particular when looking for content that is otherwise difficult to find, such as observatory publications, reports, and annals. Lists of historical scans available so far are provided by the ADS[3].

Data links: The D links available for many records allow users to access the astronomical data used in these papers. This information is typically provided by the data center or observatory where the data are available. For instance, papers using data from telescopes of the European Southern Observatory (ESO) are linked by the ESO librarians to the programs that generated the data, and the D link for

---

[3] adsabs.harvard.edu/historical.html; adsabs.harvard.edu/journals_service.html

these records will take users directly to the ESO Archive where the data can be requested.

The "Select References In" option in the Filters area lets users limit queries to papers using data from a specific observatory, for instance HST, ESO, Gemini, Subaru, and Keck.

Catching errors: Some of the content available through the ADS is provided by libraries, even though the individual links do not reveal that information. In particular, librarians supply original bibliographic records, corrections to existing (especially historical) entries, and information about data links for their specific observatory.

3.1.2   Result Lists: Bibcodes, Citation Counts, and Letter Links

ADS result lists with brief entries display a variety of information above the authors and title. These are the bibcode of the record, the citation count, publication month and year, as well as several letter links, depending on availability.

Bibcode: In astronomy, bibcodes are used to uniquely identify publications; they are assigned by the ADS. A bibcode consists of 19 characters. The first four represent the publication year, followed by up to five characters for the journal or book abbreviation, up to four digits for the volume, one optional character to indicate a special section (*e.g.*, Letters), up to four digits for the starting page, and finally the first letter of the family name of the first author. Unused characters/digits are filled with dots. A typical bibcode will be for instance 2005ApJ...635..260S, representing the article by Savaglio *et al.*, published in *ApJ*, vol. 635, starting on page 260.

Citation count: For the citation count of an article to be displayed, the respective selections in the Sorting section must be made, see above.

Publication month and year: The publication month and year are shown in the format MM/YYYY. This information is useful for instance when citations are compare, because it makes a difference how many months ago a paper was published.

Letter links: ADS entries provide links to additional information, symbolized by capital letters on results pages. Some of the information is stored on the ADS server, other links take users to services that reside elsewhere. The number and kind of links vary. The most important letter links are: A = abstract, C = citations (other articles citing the paper), D = data links, E = article fulltext in HTML, F = article fulltext in PDF, N = list of objects mentioned in the article from NED (NASA/IPAC Extragalactic Database[4]), O = associated articles (*e.g.*,

---

[4]http://nedwww.ipac.caltech.edu/

errata), R = references (articles cited by a paper), S = list of objects mentioned in the article from SIMBAD database[5], X = links to articles in the arXiv database.

The letters are usually shown in blue. Fulltext links (PDF and HTML) of subscription-based articles will only allow access to the paper if you or your institution subscribe to the journal. However, E and F links of some articles will display in green, indicating that these articles are open access, *i.e.*, available to everybody. Whether open access is granted and how soon after publication, depends on the journal. For an explanation of open access, see Section 6.

3.2 *arXiv.org e-print Archive*

The other large service in astronomy is the arXiv.org e-print repository[6]. The main server is located at the Cornell University Library. It provides open access to over half a million e-prints in physics, mathematics, computer science, quantitative biology, and statistics. For astronomers, the most important arXiv section is astro-ph, an e-print archive that dates back to April 1992. Like the ADS, also arXiv has mirror sites in various countries.

Originally developed by Paul Ginsparg when he was still affiliated with the Los Alamos National Laboratory (LANL), arXiv is now owned, operated, and mostly funded by Cornell University.

3.2.1 Manuscripts Uploaded by Authors

ArXiv/astro-ph is a collection of manuscripts posted by authors. There is no obligation for any author to make papers available, but the majority of astronomers post their papers on the e-print archive (Schwarz & Kennicutt 2004). The repository can be accessed via the web, or one can subscribe to email notifications and RSS feeds.

When submitting a paper, authors are asked to grant arXiv.org the non-exclusive right to distribute the article and certify that the work is either in the public domain or available under a Creative Commons (CC) license[7]. These licenses provide copyright holders with a mechanism for (partly or fully) waiving their rights in order to allow sharing of information, but CC license holders will still be credited if others use their works. The advantage is that obstacles typically arising from copyright laws can be avoided. Authors have to make certain that the license they grant to arXiv does not conflict with a journal agreement they may have signed[8].

To ensure that arXiv content is relevant to current research, an endorsement system was introduced in 2004. Since then, papers of first-time submitters have to be endorsed by existing members of the arXiv community before they are accepted

---

[5]http://simbad.u\discretionary-strasbg.fr/simbad
[6]http://arxiv.org
[7]http://creativecommons.org/license/
[8]http://arxiv.org/help/license

by the e-print server. In January 2009, astro-ph was split into six sub-categories. Submissions have to be assigned to at least one of these subcategories. The split is meant to make announcements of new papers more manageable for those interested only in subsets of astro-ph[9].

As many authors post their papers before acceptance by a journal, several versions of the same paper can be posted while the refereeing process continues. The author decides whether every iteration is posted as a modified version. Typically, the text of the final posted version is identical to the published paper. The layout will most probably be different as the original typeset layout is copyrighted by the publisher, and most publishers will also require that you use and cite the refereed and published version of the paper, not simply the online arXiv version.

3.2.2   Finding your Way Through arXiv/astro-ph

On the main arXiv.org screen, astro-ph is listed as the first option in the physics section. Users can browse the entire database, go directly to entries made available today or during the recent week, or search astro-ph by author name, title, or abstract words and other categories.

For each entry, the arXiv number is shown, as well as title, author, and author-defined comments, e.g., number of pages and figures, and status of the manuscript (submitted, accepted, revised version, etc.).

The arXiv papers should be cited in the format arXiv:YYMM.NNNNv# [category], for instance arXiv:0803.1234v1 [astro-ph], to indicate version 1 of paper number 1234, posted in March 2008 on the astro-ph server.

When citing manuscripts that have merely been submitted to journals and are not yet accepted, bear in mind that they are highly likely to undergo (sometimes substantial) changes before they are actually published. Indeed, scientists typically still cite the published version of papers whenever possible. Future changes may be one of the reasons, as well as the fact that the final reference reveals the journal where the paper will be published, so it conveys some of its reputation, e.g., refereed journal *versus* conference proceedings. This effect is desired when compiling a publication list for the curriculum vitae or home page.

The vast majority of astronomers post their manuscripts on astro-ph; for instance, an estimated 80% of the papers we look at in the context of the ESO Telescope Bibliography (see Sect. 9.3) are submitted to the e-print server. Schwarz & Kennicutt (2004) find that, on average, *ApJ* papers posted on astro-ph are cited more than twice as often as those not posted. Does it matter where in the daily list of new postings a manuscript is positioned? Dietrich (2008a, 2008b) finds that papers appearing at or near the top of astro-ph listings receive significantly more citations than those on lower positions. Because submissions are listed in the order they arrive at the astro-ph server, with each new list starting with papers received at 4 pm (US Eastern time), authors do have some control over the rank of their articles. It remains to be seen whether Dietrich's findings are considered

---

[9]http://arxiv.org/new/#jan2009

an unwanted drawback to the arXiv service by the managers at Cornell so that it leads to modifications in the organization of the ranking system.

4   Databases Beyond ADS and astro-ph

With ADS and astro-ph being so dominant and relevant for a large number of queries, other services are often neglected, even though they may provide additional and sometimes unique information.

4.1   SPIRES-HEP

SPIRES-HEP[10] is the High-Energy Physics database of the Stanford Physics Information Retrieval System (later renamed Stanford *Public* Information Retrieval System). It was originally developed at the Stanford Linear Accelerator Center and is now a jo- int project of SLAC, DESY (Deutsches Elektronen-Synchrotron), and FNAL (Fermi National Accelerator Laboratory, Fermilab), managed by the SLAC library.

SPIRES-HEP provides bibliographic information about high-energy physics literature, including journal articles, theses, reports, conference papers, books, and others, with links to the fulltexts of these items. Among others, arXiv is one of the sources from where fulltexts of papers referenced in SPIRES-HEP can be obtained.

4.2   Google Scholar

Google Scholar (GS)[11] is a free research tool for scholarly literature and citations. While it is known that Google Scholar covers journal articles as well as conference proceedings, books, reports, and other media, and that even full-text repositories are searched, GS keeps users in the dark about exactly which sources it is using.

Similar to the main Google search engine, Google Scholar also does a very good job of correcting misspellings and suggesting different spellings including name variations. It covers a wide range of subjects, but seems to be generally stronger in the fields of social sciences, arts and humanities, and engineering (see for instance Harzing & van der Wal 2008). One drawback is that result lists can contain duplicates.

4.3   Scirus

Another literature search tool is Scirus[12]. It is maintained and provided free of charge by Elsevier Science, one of the largest scientific publishers. As of January

---

[10]http://www.slac.stanford.edu/spires/hep/

[11]http://scholar.google.com

[12]http://www.scirus.com

2009, over 480 million items are indexed, including refereed and unrefereed publications, repositories, science-related websites from the American Physical Society (APS), arXiv, the Institute of Physics Publishing (IOPP), NASA, and many other sources. Scirus offers a sophisticated advanced search interface that allows users to filter by date range, information type, and subject area (including astronomy). However, this tool only helps for retrieving publications and other information resources; citations are not provided.

4.4   ISI Web of Knowledge and Web of Science (WoS)

The ISI (Institute for Scientific Information) Web of Knowledge[13] is a platform that provides access to a variety of databases (*e.g.*, Inspec, Medline, Biosis) and tools for retrieving and analyzing research articles. Some backfiles, including citations, date back as far as 1900.

The citation database included in the ISI Web of Knowledge is called Web of Science (WoS)[14]. It covers over 9000 publications in the sciences, social sciences, and arts and humanities, but is limited to journals, so that citations from books, reports, and other publications not included in the WoS are not counted. WoS is used by many organizations to analyze their scientists' research output, but it has been repeatedly criticized for omitting citations from books and for focusing mostly on English-language titles from the US and Europe.

In the past, the Journal Citation Report (JCR), which is prepared annually by the Web of Knowledge and which presents journal rankings, has been criticized for reporting flawed journal impact factors because of difficulties in handling possibly non-unique journal abbreviations. In astronomy, the *Astrophysical Journal* and *Astronomy & Astrophysics* have been particularly affected by this mistake (Sandqvist 2004). In the meantime, the error has been corrected and the acccurate journal impact factors published.

Unlike SPIRES, Google Scholar, and Scirus, WoS is not free. It is a commercial system, marketed by Thomson Reuters, that needs a subscription.

4.5   Scopus

Described as the "largest abstract and citation database of research literature and quality web sources" on their web site, Scopus[15] has become a serious competitor for WoS. As is the case with WoS, Scopus requires a subscription. The tool was developed and is marketed by science publisher Elsevier.

Scopus offers information from over 16 000 peer-reviewed journals, as well as conference proceedings, book series, scientific web pages, and repositories, including arXiv. Full-text articles can be accessed seamlessly, provided the necessary subscriptions are maintained. Scopus aims for complete coverage of their records

---

[13]http://www.isiwebofknowledge.com
[14]http://isiwebofknowledge.com/products_tools/multidisciplinary/webofscience/
[15]http://www.scopus.com

from 1996 onwards, but historical material has also been added to the system, going back as far as 1823 in some cases. Citation analysis is provided from 1996 onwards.

One strength of Scopus are the innovative tools for analyzing publication and citation statistics, including the h-index and unique graphs.

The scope, coverage, and approach of Google Scholar, WoS, and Scopus are quite different, with WoS and Scopus providing substantial factual information, various search and ranking features, help pages, and guidance on how to use the services best. They act as an information hub (Jasco 2005). In contrast, Google Scholar applies the Google PageRank algorithm to display search results, which typically leads to the most popular hits being displayed at the top. Advanced search and display options are limited in comparison with WoS and Scopus.

Differences are becoming even more obvious when looking at citations. Meho & Yang (2007) find that the overlap in citations between Scopus and WoS is rather small, namely only 58%, and only 31% between Google Scholar and the union of WoS and Scopus. This emphasizes that none of these databases provides an absolute answer regarding completeness of search results and, specifically, numbers of citations. The three services complement each other, but none of them should be regarded as the perfect search system.

The main advice clearly is: when using publication and citation databases, never switch off your own thinking. Always evaluate all resources and put them in a context with other results.

5   Core Astronomy Journals

Every science has journals that are essential to that discipline. In astronomy, the number of these most important journals is small: only six titles are typically referred to as the "core journals." These are *Astronomy & Astrophysics (A&A), Astronomical Journal (AJ), Astrophysical Journal* and its *Supplement Series (ApJ/ApJS), Monthly Notices of the Royal Astronomical Society (MNRAS)*, and the *Publications of the Astronomical Society of the Pacific (PASP)*. These journals alone produced on average 36% of all refereed journal articles listed in the ADS for the years 2000–2007.

All core journals in astronomy are published on behalf of organizations or learned societies. *A&A* is published on behalf of the European Southern Observatory (ESO), *AJ* and *ApJ/ApJS* by the American Astronomical Society (AAS), the *Monthly Notices* on behalf of the Royal Astronomical Society (RAS), and *PASP* by the Astronomical Society of the Pacific (ASP). This means that, even if the journals are produced and marketed by commercial publishers, their (partly understandable) economical interests are somewhat leveled out by the societies who often influence or even control the subscription price.

The core astronomy journals are subscription-based; *i.e.*, the latest issues are only accessible with a valid subscription. However, all of them apply the "delayed open access" model. After two or three years, articles are made available to everybody (for a description of open access, see Sect. 6).

All of them are currently still available in print, as well as electronic format, even though some publishing models are already changing. For instance, the default subscription of the *ApJ Letters* includes electronic-only access as of 2009, and also the basic subscription price for *A&A* will soon only cover the electronic version; in both cases, the print version needs to be explicitly ordered (and paid for). These are first moves towards e-only publishing in astronomy, and it will only be a matter of time when the other core journals follow.

While librarians are not directly involved in the production of the journals, they pay the subscription fee, negotiate licenses (which define who can access the publication, and under which conditions), make journals available through web pages and library catalogs, and troubleshoot in case of access problems.

A&A, AJ, ApJ/S, MNRAS, and *PASP* are available through individual subscriptions that do not require customers to purchase a whole package of journals. Content from all titles can be found through the ADS or via astro-ph, provided that the authors submitted their manuscripts to the arXiv e-print server. In addition to these well-established finding tools, publishers are also starting to develop their own platforms to offer easy access to all their journals, plus additional information like commenting systems, readers' communities, etc. For instance, users of IOPscience[16] will be able to search latest and historical research articles published by IOP. Finding tools range from the established author/title/keyword searches to emerging technologies, such as tag clouds. Retrieved content also includes e-prints from arXiv (presented via an IOP-developed e-print interface) and certain web pages.

This trend away from static journals towards "virtual journals" or article databases can be expected to grow in future. It remains to be seen whether publishers' services will be able to compete with the "master search engine" provided by the ADS.

6   Open Access

6.1   The OA Concept

A brief description of open access publishing is provided by Wikipedia: "publication of material in such a way that it is available to all potential users without financial or other barriers"[17].

The open access movement goes back to the 1960s, but has gained momentum mostly since the last decade of the past century. Among other reasons, it was prompted by the constantly rising subscription prices of scholarly journals that has led researchers, librarians, funding agencies, publishers. and others into discussion about how scientific publications can be made available in a better way. The main focus is on peer-reviewed, publicly funded literature. The idea is to remove some of the existing barriers, most notably price barriers (subscriptions, licensing fees,

---

[16]http://iopscience.iop.org
[17]http://en.wikipedia.org/wiki/Open_access_publishing, accessed Jan. 31, 2009

pay-per-view charges) and permission barriers (copyright and licensing restrictions, Suber 2007) and make them accessible to everybody. Of course "everybody" is a relative term, as some impediments continue to exist, for instance access limited to members affiliated with an institution, language barriers, accessibility barriers, etc.

There are various forms of open access:

- Self-archiving on authors' home pages: author self-archiving is often called the "green road" to OA. Most publishers (also commercial ones) grant their authors the right to make manuscripts, including the final version, available on the author's web page. If the final typeset version may be used, the publishers typically demand that the bibliographic details are given so that readers are pointed to the publisher's journal. The SHERPA RoMEO website[18] lists publishers' copyright and self-archiving policies.

- Self-archiving in a repository: these can be institutional or subject-based repositories. Many institutions have set up repositories where all affiliated researchers deposit their articles. Some scientific disciplines maintain central subject-based repositories. In astronomy, such a repository is the astro-ph e-print server.

  Self-archiving by scientists neglects two very important aspects: retrieval via a central access point, and preservation. Repositories are more reliable in this regard.

- Publishing in OA journals: typically called "open access publishing", this refers to journals that make their content available without subscription. In today's publishing landscape, many different flavors of OA are applied by publishers, ranging from delayed access (with free access to content after a given time) to institutional membership (whereby all papers are immediately available for those affiliated with the member institute), and on to the "author pays" model (*i.e.*, authors pay a certain fee to the journal, readers don't).

Even though removing barriers related to (sometimes unaffordably high) prices is one of the key incentives of the open access movement, it must be stated that serious OA advocates do not claim that the *production* of open access publications is without costs. The production of OA journals may overall be less expensive than conventionally published literature; however, it still demands funding that needs to be provided by someone. The word "free" (of charge) can therefore only refer to the readers of open access publications, not to the publication process at large.

One of the advantages open access provides for authors is better accessibility because, theoretically, every interested researcher can obtain the publications. Whether or not this actually leads to more citations is still being investigated. For

---

[18]http://www.sherpa.ac.uk/romeo.php

a discussion of this topic in the case of *The Astrophysical Journal*, see Kurtz & Henneken (2007).

According to the key findings of the third ALPSP (Association of Learned and Professional Society Publishers) survey of Scholarly Publishing Practice[19], the proportion of publishers that offer optional open access (authors pay a fee) has grown from 9% in 2005 to 30% in 2008. However, the report concludes that "the take-up of the author pays open access option is exceedingly low." This statement is not surprising, given that (i) the fee authors have to pay for optional OA is sometimes as high as 3000 Euros and (ii) that, especially in sciences like astronomy the majority of research literature is available either from the arXiv e-print server or directly from the journals through "delayed open access." As a result, astronomers may not be particularly attracted by a publishing model that requires further payments from them.

A description of the state of open access in astronomy can be found in Grothkopf & Erdmann (2008).

6.2 OA Finding Aids

Several finding aids can be recommended for retrieving open access publications.

6.2.1 Directory of Open Access Journals (DOAJ)

The idea of creating a Directory of Open Access Journals[20] goes back to the First Nordic Conference on Scholarly Communication held in Lund and Copenhagen in 2002[21]. The DOAJ Head Office is located at the Lund University Libraries. As of January 2009, the directory includes over 3800 scientific and scholarly journals, of which more than 1300 are searchable at article level. Open access journals are defined as peer-reviewed journals that "use a funding model that does not charge readers or their institutions for access" and that make all content freely available without delay.

6.2.2 OAIster

OAIster[22] is a union catalog of digital resources, compiled by harvesting descriptive metadata using the Open Archives Initiative Protocol for Metadata Harvesting (OAI-PMH). With a current staff of only three, the OAIster team has put together an impressive catalog of digital resources, including digitized books and articles, original ("born-digital") texts, audio files, images, movies, and datasets. These items are often located in the so-called deep web, hidden from search engines by firewalls. OAIster reveals these resources by using OAI-PMH and makes them

---

[19]http://www.alpsp.org/ngen_public/article.asp?pfs=0\&did=47\&aid=27749\&oaid=-1
[20]http://www.doaj.org
[21]http://www.lub.lu.se/ncsc2002/
[22]http://www.oaister.org

available via a searchable interface. In December 2008, OAIster provided access to more than 19 million records from over 1000 data providers.

6.2.3   ScientificCommons

The aim of the ScientificCommons project[23] is to "provide the most comprehensive and freely available access to scientific knowledge on the internet". Like OAIster, ScientificCommons uses the OAI-PMH protocol to retrieve information from repositories. According to their website, ScientificCommons had indexed more than 13 million scientific publications as of January 2007. Author names are extracted across institutions and archives so that distributed publications can be found via one common interface. ScientificCommons also extracts professional relations between authors to make their research development transparent to the public. Personalization services include an option for authors to organize their publications directly with ScientificCommons.org in order to create their own researcher profiles.

ScientificCommons is a project of the University of St. Gallen (Switzerland), and is hosted and developed at the Institute for Media and Communications Management.

6.2.4   Directory of Open Access Repositories (OpenDOAR)

OpenDOAR[24] is a compilation of information about and links to academic open access repositories. It aims at increasing their visibility and retrievability by providing end-users with a search interface to locate specific archives by geographic location, subject, content type, and more. Physics and Astronomy are currently represented with 45 repositories.

From its start at the end of 2005, the size of OpenDOAR has increased quite steeply to reach its current coverage of approx. 1300 repositories. Almost half of the included repositories are located in Europe, approx. 30% in North America, and around 20% in Asia, Australasia, and South America combined. OpenDOAR is maintained by the Unversity of Nottingham in the UK.

7   Problems with Online Documents

The internet and the advent of electronic publications have enormously modified how scientists use the literature. Especially in astronomy, we are used to finding (seemingly) all research publications on the internet in electronic format. The ease of using online publications and their immediate availability should not let us forget that there are some severe obstacles on the way to the e-only scholarly society.

---

[23]http://www.ScientificCommons.org
[24]http://www.opendoar.org

7.1 Long-Term Access

Ever since the advent of electronic publications, librarians have communicated the challenges in long-term access to the research community. Two major topics have always been of concern: archiving and preservation.

Archiving refers to questions such as:

- what material (scientific articles, electronic communication, databases, etc.) and which resources should be archived? (web pages only from certified providers? "everything"?)

- who should be in charge of archiving? (individual data providers? national libraries and archives? publishers?)

- where should the archived content be stored (private web pages, libraries, data centers, commercial archive providers, etc.) and how many mirror sites should exist?

Preservation of digital information resources encompasses assuring global perpetual access; provision of tools to read and interpret the data; migrating storage media to newer technologies, including all associated metadata, access rights, provenance information and, most importantly, the links within the document and those reaching out to other media items. Reference systems must be in place to make certain that the information can be found. The authenticity (identity) of electronic publications needs to be easily recognizable in order to determine which version of the document is being used, and their links and connections with other digital objects must be transferred to each newer storage medium. Finally, mechanisms are needed to guarantee the integrity (intactness) of digital information.

7.2 Completeness of Digitized Material

In order to make historical literature, *i.e.*, material originally published on paper, available online, numerous scanning projects are underway at observatories, universities, the ADS, and other data providers. How complete are the results? At the minimum, all projects scan the scientific content, *i.e.*, the articles published in the journals, magazines, conference proceedings, books, or other items that are being digitized. But what happens to other parts, such as errata, news sections, advertisements, letters to the editor, front and back matter, obituaries? Depending on the treatment received during and after digitization, such material may be lost or destroyed.

7.3 Deleted Items

In the print environment, the path of publications is quite well-defined. After production at the publisher, it is sent to libraries where the material is displayed and later bound into volumes and placed on shelves, available for future access whenever needed. Nothing gets lost or is changed, the scholarly content is archived the way it was originally published.

In the digital environment, things are more short-lived. For instance, what happens if a journal publisher finds out that an article plagiarized someone else's work, or if the content, if acted upon, poses a serious health risk? In some cases, it might be considered necessary to remove the article from the publisher's website, even though it had been formally published (see for instance the Elsevier policy on article withdrawal[25]).

Other examples are Wikipedia pages of ceased projects or comments and contributions posted to mailing lists and blogs that may be removed. While such items are intrinsically more ephemeral, they show that the availability of online content should not be taken for granted.

7.4   Broken Links

In today's use of web resources, it often seems to be sufficient to indicate specific pages through Uniform Resource Locators (URLs). While many of them reveal which organization or company hosts the source (*e.g.*, en.wikipdia.org) or what specific subdirectories are on that server (*e.g.*, .../wiki/Url), they can also simply consist of figures and then not tell users anything (at least not at first sight) about their origin and host.

The real danger, however, is that URLs point to one exact location. Should the location change or any part of the string be modified, users will not be able to retrieve the intended web page, but will be sent to an error page.

To avoid such broken links, a better naming system has been developed: Digital Object Identifiers (DOIs[26]). DOIs are names, not locations, and provide a unique, permanent digital identifier to objects. Mostly used for scientific publications, it has been widely applied by publishers. By the end of 2008, approx. 40 million DOI names had been assigned.

A DOI follows a defined structure and may look like this: 10.1065/abc123defg, where 10.1065 is the prefix. The first two digits are the directory code; 10 indicates the publishing sector. 1065 represents a given publisher; abc123defg is the suffix. It is chosen by the publisher and has to be unique within the realm of the prefix. DOIs can be resolved through browsers, for instance at the website dx.doi.org. The site name and the DOI are seperated by a forward slash[27].

The system is managed by a membership consortium called the International DOI Foundation. The official link registration agency for scholarly and professional publications is CrossRef[28], a citation-linking network that covers millions of articles and other content items. In order to have DOIs assigned to their publications, publishers have to become CrossRef members and pay an annual fee.

---

[25]http://www.elsevier.com/wps/find/intro.cws_home/Article\Withdrawal
[26]http://www.doi.org
[27]for instance, http://dx.doi.org/10.1065/abc123defg
[28]http://www.crossref.org/

8 Everything is on the Internet ... Really?

Compared to other sciences, astronomy traditionally has been on the forefront of the move towards electronic publications. In 1997, only two years after the start of electronic publishing, almost half of the peer-reviewed literature in astronomy was available in electronic format (Boyce 1998). This may lead to believing that by now, *everything* has been made accessible online. This assumption is understandable, but wrong. Information resources that are not available on the internet often are forgotten, neglected, ignored.

Information not easy to find includes material such as:

- special supplements accompanying (in particular historic) publications, such as photographic plates;

- literature in radio astronomy; this refers in particular to historical publications, but also to recent literature published in engineering journals;

- observatory publications, especially those older than two or three years;

- books and conference proceedings, in particular if published more than approx. five years ago;

- non-English language literature.

When looking for literature, it is crucial to use various retrieval tools and not rely on "quick & dirty" internet searches that may tempt you to believe that the result is already complete, while in reality it is not.

9 Organizing a Personal Library

Having access to the literature is one part of using information, but keeping your own library of papers that are of interest may be referred to in your own publications is another. An additional difficulty arises from journals typically applying their own style for citations. This means that authors need to follow certain guidelines when they compile reference lists for their manuscripts. These guidelines are different for each journal.

For many years, various systems have been around to help scientists with this task. These were mostly commercial programs that had to be purchased and installed on a computer, so-called reference management software. Such software stores information about (scientific) articles, and the bibliographic citations can then be output in various formats to be used in reference lists.

More recently, open source software systems have been made available that are available to scientists at no cost or for a small fee. The majority of these programs are still used locally on the user's desktop, but some have a web complement or are even entirely web-based.

Standard features that most reference management systems fulfill include:

- create, import, export records;

- assign notes, keywords, or tags;

- create collections for specific topics or purposes;

- store copies of the articles (in PDF format) for off-line reading.

Of special interest here is in which format records can be exported. Astronomers usually compile reference lists in one of the following two ways.

- LaTeX: in many cases the list of references is the last section of a manuscript. References are manually added as soon as they are cited in the text. The LaTeX reference list is directly integrated in the manuscript

- BibTex: a BibTex reference list (.bib file) requires a bibliography style file outside of the manuscript which calls the reference that is cited in the manuscript.

Many reference management systems are able to handle BibTex references (*e.g.*, export records in BibTex format), while only a few have a direct link (export feature) for LaTeX references. If references need to be exported from systems without direct LaTeX integration, it is recommended to export them first to a "plain" export format (*e.g.*, a clipboard) and copy and paste them from there into the LaTeX reference list. A good comparison of reference management software can be found in Wikipedia[29].

In the following, we briefly explain four reference management systems and outline some specific characteristics. These four systems are JabRef, Papers, Zotero, and Mendeley.

- **JabRef:** The JabRef[30] reference manager is available as freeware under the GNU General Public License. It is platform-independent so that it runs equally well on Windows, Linux, and Mac OS. JabRef's main features include the ability to

  − import, organize, annotate, and archive

  records. Existing BibTex libraries (.bib files) can be imported, and the developers claim that it is also possible to fetch entries from the arXiv eprint website. PDFs can be read offline so that users are independent of internet access. Exporting records requires custom export filters, *i.e.*, layout files that reside outside of JabRef. These must be written by users and can be

---

[29]http://en.wikipedia.org/wiki/Comparison_of_reference_management_software
[30]http://jabref.sourceforge.net/

shared via SourceForge.net. JabRef is built around BibTex and is a good tool for users who are comfortable using BibTex.

Users can create various collections, for instance, to group records by topic or to keep all records pertaining to specific reference lists in one place. In addition, individual records can be highlighted for easier retrieval.

- **Papers:** Papers[31] is a reference management software that has been developed specifically for the Apple Macintosh operating system. It is not free of charge; a single-user license costs EUR 29 as of mid-2009. Papers is integrated with the iPhone so that articles can be synchronized.

  The software allows users to

  - find, import, organize, annotate, and archive

  libraries. As is the case with JabRef, import plug-ins exist for BibTex files and other formats. A specialty of Papers are several built-in search engines that provide direct access to articles databases, including the ADS and arXiv. Without having to leave Papers, users can search the repositories, save repeating searches, and use some of their search features, like limiting them to first author searches or other specific fields. Retrieved records can be imported with one keystroke. The search history is accessible from within Papers so that searches can be re-used easily.

  Papers intends to encourage users to read articles on-screen by showing PDFs in full-screen mode, providing the opportunity to annotate them, and by remembering the last reading position in the file. In addition to bibliographic information about articles, Papers also allows users to store email addresses of author names and passwords for access to fulltexts of journals.

  Exporters exist for BibTex, Word2008, and CSV (comma-separated values) formats. Unfortunately, the system does not provide an option to add keywords to records or tag them, which is quite surprising for a software with otherwise high functionality.

- **Zotero:** Zotero[32] is a plug-in for the FIrefox 3.0 (or higher) browser. It can be used on Windows, Mac, or Linux machines. The main features include options to

  - find, download, organize, tag, and archive

  records. Existing libraries can be imported from intermediate formats, *e.g.*, BibTex. The main specialty of Zotero, however, is the ease with which records can be captured directly from the web. Simply by clicking on the little notepad or paper icon in the URL-field, web page content can be imported

---

[31]http://mekentosj.com/papers/
[32]http://www.zotero.org

and then be processed further, *e.g.* by assigning records to collections that have been created or by attaching files (PDFs, screenshots, etc.) to records. Related papers can be linked to each other.

Libraries (collections) can be exported for instance to BibTex or Rich Text Format. Alternatively, a bibliography of selected records can be created using any of the journal styles included in the large style gallery. The second option will provide records formatted according to that particular journal's citation style. Unfortunately, no astronomy-related journals are currently included in the list of journals, so that astronomers will probably simply end up creating a bibliography of the records they want to export, copy it to the so-called clipboard, and then format it manually.

The Zotero window can be collapsed entirely in the Firefox browser, in which case only the Zotero icon is visible in the lower right corner. When in use, the program will either occupy the lower third of the browser window or it can be opened to fill the whole window. Zotero is organized in three columns, with the left one holding the various collections, the middle one titles, authors, and publication years of the individual papers, and the right column showing details of the currently edited record. Article PDFs can be stored with the records for offline use, and users can assign tags and notes to records. In addition to published literature, Zotero can also take snapshots of web pages, which can then be viewed offline, annotated, and marked using a highlighter tool or little post-it stickers.

In order to visualize records in the collections, Zotero offers an option of viewing a timeline. Timelines are organized on three levels; filters can be assigned to limit the number of records shown, and highlighting specific keywords allows users to view publishing trends.

- **Mendeley:** Mendeley[33] is a reference management tool for the desktop (Windows, Mac, Linux) with an online counterpart. The program is free of charge, but users need to sign up in order to download the software. The current version (as of mid-2010) is still beta (0.9), indicating that Mendeley is still being developed. The program intends to provide a system with which users can

  – manage, share, and discover

research papers. Records can be imported directly from the ADS, arXiv, Google Scholar, and other databases. Exporting records from Mendeley is currently possible in Word and OpenOffice Writer, using more than 100 citation styles. Once again, astronomy journal styles are not available.

A special feature of Mendeley is its ability to automatically extract information from PDFs through drag and drop. From files dropped into Mendeley,

---

[33]http://www.mendeley.com

the system will import bibliographic details along with DOIs, arXiv IDs, and other information. PDFs can be annotated using a highlighter tool or sticky notes. Mendeley's powerful full-text search capability retrieves search terms not only in author names, titles, and keywords, but in the entire text of articles.

The program is very much geared towards establishing collaborations among colleagues and tries to act as a social networking tool. By using the web part of the system, users can form networks and invite others to share their papers and discover new ones. Research trends are visualized, for instance, by displaying the most popular authors and topics and through statistics about the papers in the collection.

All systems described above have some advantages and some disadvantages. JabRef is platform-independent, but its use is not very intuitive, because the program is built around BibTex and therefore will be used mostly by those who are familiar with BibTex. The second program, Papers is more user-friendly, and it searches the ADS and arXiv, but is only available for the Mac and is not available entirely free of charge. Zotero, the Firefox plug-in, conveniently captures metadata directly from the web, thus allowing easy import of records of articles and web pages, but the style gallery for exporting records is not very suitable for astronomers so that manual formatting of references is necessary. Mendeley provides a remarkable full-text search capability aimed at building communities among scientists. Right now, the functionality is still somewhat limited, but the developers seem to be moving fast in order to incorporate more features.

In the end, the decision to use a particular reference management system will depend on personal preferences and on which system matches most of a user's selection criteria.

10 Bibliometric Studies

Publicly funded organizations typically have to report back to their management and funding agencies about how their resources were spent and whether they provide the expected output. For scientific entities, the output is increasingly measured by bibliometric methods. Often librarians are in charge of these studies.

Bibliometrics can be defined as a set of methods used for publication and citation analysis to explore the impact in the respective field. Observatories tend to use such metrics to evaluate the research output based on data from their telescopes and instruments, and the acceptance of these facilities among the user community.

Some commonly used bibliometric methods are listed in Table 1. All of them have both advantages and disadvantages. For instance, counting the number of publications reveals the productivity of an entity, but gives no information about the impact and importance of these publications. Citation analysis, assuming that important articles are cited more often by other authors, does reflect the impact among peers. However, citation counts can be inflated, for instance by

Table 1. Advantages and disadvantages of metrics.

	Good	Bad
# Publications	Productivity	No impact
# Citations	Impact	Can be inflated
Mean/median cites	Allow comparison of	Reward low
per paper	different eras	productivity
"High-Impact Papers"	Show trends	Favor "hot topics"
h-index	Productivity + impact	Determined by
		years of operation

erroneously counting citations of different scientists who have the same name and initials, because of friends and colleagues who systematically cite each other, or even because of articles that only point out incorrect findings in the cited work (Meho 2007). Calculating the mean or median number of citations per paper has the advantage of allowing comparison (bearing in mind the difficulties intrinsic to such comparisons, see Sect. 9.1) of publications from different scientists or from various facilities, but high citations to only a few articles can increase the average value, which may wrongly suggest that the majority of papers included in the result set have received high cites.

Another measure are so-called High-Impact Papers (HIPs), a term coined originally by ISI (now Thomson Reuters) and later applied to publications in astronomy (Meylan *et al.* 2004). An HIP is defined as a paper that belongs to the 200 highest cited refereed papers in a given year. Once this subset is defined, *e.g.*, by using the ADS Abstract Service, papers using observational data are analyzed for the observing facilities that provided the data. This method shows interesting trends, in particular if applied over several years. However, it is very time-intensive to compile and may favor "hot topics" that receive a lot of momentary attention.

A few years ago, Hirsch (2005) proposed a new method, the so-called h-index. It aims at quantifying a researcher's scientific research output in a more balanced way by combining numbers of publications and numbers of citations, hence productivity and impact. The idea seems to be simple and yet more effective than many other metrics: the h-index is defined as the number of publications that have received at least h citations. Using the ADS or other citation databases (*e.g.*, Web of Science, Scopus), the value can be computed quite easily. First, submit a query to retrieve all papers for which you would like to calculate the h-index. For instance, such a query can be for all publications by a specific author, using data from a particular telescope, or all scientists affiliated with a specific organization. The set of results needs to be ordered by decreasing number of citations, so that the paper with the highest number of citations resides at the top of the list, the one with the lowest citation number at the bottom. At that point, h can be found where the rank number of the paper (starting with rank number 1 at the top) is higher than or equal to the number of citations the paper has received.

As might be suspected, h computed as described cannot be the same for authors that have been publishing for many years and for those who just entered research; naturally, the senior researcher will have published many more articles that will have gathered many citations in the course of the years. Similarly, h will also not be comparable across disciplines or even across subfields of disciplines; *e.g.*, papers published in the subject area of planetary astronomy will on average have different h-indices than, for instance, UV astronomy publications. To make comparisons possible, at least to some extent, Hirsch introduced a second measure, the so-called m-parameter, which is calculated by dividing h by the number of years. For instance, this can be the number of years of scientific activity of researchers or the number of years of operation of facilities: $m = h/t$.

10.1 Caveats of Citation Analysis

Regardless of which bibliometric method is chosen, there are some general problems one should keep in mind.

- *Incompleteness:* by now, several web-based services have been developed that provide citation information. Some of them (*e.g.*, Web of Science, Scopus) require a subscription, others (*e.g.*, ADS Abstract Service, Google Scholar) are free-of-charge. The sources and technologies the services use to obtain and analyze the data vary, and citation statistics can be vastly different. Therefore, it cannot be emphasized enough that one single citation database will never give the full picture.

- *Incorrectness:* the sources of incorrect citation information can be manifold. A simple, but frequent, one is that authors make mistakes in their lists of references. Accordingly, these references cannot be resolved by the citation databases and are not assigned to the intended cited paper, *i.e.*, the citation will be ignored. Another reason for errors lies in different abbreviations used for the same journal. References to each of them may then be assigned to seemingly different journals, and citations may be counted in parallel rather than cumulatively.

- *Citing behavior:* it should be remembered that many authors prefer to cite well-established scientists, those that are cited by everybody. There may be many reasons for this behavior: it may arise from a wish to "go with the flow" of what other authors do; it may save time to use reference lists from papers written previously; authors may have a wish to show that they read the "classics" in the field, or perhaps they want to show "solidarity" with friends and close colleagues. Regardless of the exact motivation, this creates a bias towards highly cited scientists and puts young and less connected authors at a disadvantage.

- *Multi-author papers:* authors who publish in large collaborations[34] by definition create a professional network of many colleagues who will (presumably) cite this paper in their future publications. Obviously, this leads to a much larger basis of potential quotes than papers published by single authors or small groups of co-authors will have. A way to handle references to such many-authored papers would be to assign only normalized (proportional) citation credits, *i.e.*, 1 divided by the number of authors.

One more word of caution. It is very bad style to quote without actually having read the article, even though everyone in the field seems to quote it. One also would risk attributing something to the cited author that he or she did not actually write.

10.2   Alternative Measures

Citation analysis is an important part of bibliometrics, but it has one severe disadvantage: citations do not measure actual use: *i.e.*, how often were papers consulted with or without subsequent inclusion in the final list of references?

A new measure to better reflect the use of publications has been introduced by the ADS. Instead of citations, readership information is being collected. A *read* is counted as each time a user requests information via one of the ADS letter links. In 2002, 50% of the requested items were abstracts, followed by one of the fulltext versions (38%) and citation lists (8%) (Kurtz *et al.* 2003). Using their database of logs, the ADS team can identify relations between the private act of reading a paper and the public one of actually citing it. The reads history is available for ADS entries from the record's abstract page.

A similar approach is pursued by Citebase[35]. This experimental citation database looks at *downloads* of papers from arXiv.org. As of January 2009, only the UK-site is monitored, but the other arXiv mirror sites are ignored, including the main server in the U.S. This means that the number of downloads per e-print displayed is notoriously too low. It will be interesting to follow the development of this project.

On a more general level, access and download statistics can also help for evaluating the use of journals and their overall acceptance within the community of scientists. Libraries occasionally consult such statistics to determine which journal subscription should be continued and which may no longer be necessary.

10.3   Telescope Bibliographies

At many observatories, librarians are in charge of maintaining databases of papers written by staff or other astronomers that use data from the observatory's

---

[34]In a recent paper, Crabtree (2008) found that the average number of authors has increased from 2.5 in 1980 to almost 7 in 2006. In the context of the ESO Telescope Bibliography, we noticed a 2008 Nature paper with no fewer than 93 authors.

[35]http://www.citebase.org

facilities. These databases are typically compiled by scanning the important astronomical literature (manually or semi-automatically), trying to locate references to instruments and telescopes that were used to gather the data.

There are many reasons for maintaining telescope bibliographies, for instance, to measure the scientific impact of telescopes and instruments. The performance of existing facilities can also provide important guidelines for future facilities. Funding authorities often request bibliometric studies because they help them to understand the overall productivity and impact of the funded organization.

Following the observatory's policies for acknowledging telescope use is also important for the authors. Some funding agencies demand that their support be acknowledged, and also many observatories ask (or even oblige) authors to mention use of the facilities in a footnote[36]. Mentioning facilities in the footnote will also help compilers of telescope bibliographies to keep track of the scientific papers published with their data and may influence observing proposal committees and other groups for granting future observing time. Last, not least, publishing with data from renowned facilities is also advantageous for authors as it increases their visibility and helps them to be recognized by colleagues.

A list of telescope bibliographies from major ground-based and space-based observatories is available from the ESO library[37]. In addition, the ADS provides access to several keyword lists, among them many observatories. On the main ADS search screen, scroll down to the Filters section and choose "Select References In / All of the following groups" and select the observatory of interest. Note, however, that the selection criteria, the years covered, and the range of completeness will differ. To find out the exact coverage, click on the facility's name, which will take you to a short explanation. As explained before, the D link of ADS records will lead to the data center or archive from where the data used in the respective paper can be retrieved.

11   Cooperations of Librarians and Astronomers

Traditionally, librarians have undertaken the role of mediator between information providers and scientists. Many librarians regularly inform the astronomers at their institute about new developments from publishers and database hosts, and report back any concerns, questions, as well as suggestions for improvements. You will know the librarian at your home institute, and if you are looking for a librarian at another institution, the Directory of Astronomy Librarians and Libraries[38] may be of use.

On a larger scale, there are some more formal ways to enhance communication between the various groups, in particular, an IAU Working Group and the LISA conferences.

---

[36] For instance, the policy for publishing with ESO data can be found at
http://www.eso.org/sci/observing/policies/publications.html
[37] http://www.eso.org/libraries/publicationlists.html
[38] http://www.eso.org/libraries/addresses/addresses.html

11.1   IAU Commission 5: Working Group Libraries (WG Lib)

The Working Group Libraries is associated with the IAU commission 5 "Documentation and Astronomical Data." It aims at better cooperation between astronomers and librarians. The working group was officially recognized for the first time in 1990, but librarians were involved in Commission 5 activities even before that. Information about past and present activities can be found in the *Reports on Astronomy* (published every three years in the *Transactions of the IAU*), as well as on the working group's web pages[39].

During IAU General Assemblies, the Working Group often hosts so-called business meetings. These sessions are open to everybody interested and typically focus on data access and preservation, science metrics, and the role of librarians in fostering research.

11.2   Library and Information Services in Astronomy (LISA)

In 1988, the first conference on Library and Information Services in Astronomy was held in Washington, DC, at the U.S. Naval Observatory. Since then, LISA has been developed into a series of conferences, and four further meetings took place, each attended by more than 100 participants. LISA conferences provide a unique meeting point for librarians, astronomers, publishers, and computer specialists from around the world. The wide geographic distribution of participants is indeed one of its main characteristics.

Thanks to the relatively small number of core journals, in many cases coupled with quite generous funding, astronomy libraries have often had access to evolving technologies earlier than other subject areas, and have applied them early on in their day-to-day work. In addition to changing technology, the professional role of librarians has undergone considerable development. This is reflected in the topics covered at conferences. Today, LISA is a forum to exchange experiences, view topics from different angles, and gather information about emerging fields of interest in astronomy libraries.

A description of the history of LISA meetings, their logistics, and topics can be found in Corbin & Grothkopf (2006); a web page provides links to LISA conferences, as well as proceedings of past meetings[40].

12   Conclusions

The publication paradigm continues to shift from printed material to electronic formats for scientific literature, leading to an evolved concept of library services and information access. Librarians are taking on new and diversified roles. Observatory librarians fulfill the information needs of astronomers and engineers and develop digital services to give greater access to scientific content. They support their

---

[39]http://www.eso.org/libraries/IAU-WGLib/
[40]http://www.eso.org/libraries/list.html

library users when conducting research and provide help so that existing tools can be used efficiently.

In this chapter, we have introduced the varied research assistance libraries provide. Some tips and tricks were provided to help especially young astronomers to retrieve information from the ADS digital library and the arXiv (astro-ph) e-print repository. An overview was given of the most important astronomy journals and the concept of open access (OA) publishing. Because of the widespread habit of astronomers of posting manuscripts on astro-ph, OA affects astronomy and astrophysics less than many other subject areas, yet it should not be assumed that nothing remains to be done.

Scientists have gotten very used to the idea that all necessary information is available (typically free-of-charge) on the internet. However, there are some severe problems regarding long-term access, completeness of digitized material, and broken links, to name just a few. Librarians and information providers are working towards more reliable systems, but it is essential that scientists are also aware of these problems.

Increasingly, librarians maintain telescope bibliographies, *i.e.*, databases of articles that use observational data. From these databases, publication and citation statistics are derived to provide information about the scientific output of specific facilities to management and funding organizations. Telescope bibliographies conclude the "life cycle" of observations because they close the loop from observing proposals to observations and to the published results, and from there back to the underlying data.

Finally, we have shown examples of close cooperation between librarians and astronomers that increase communication and, ultimately, lead to more efficient research.

This research has made extensive use of the NASA Astrophysics Data System; it is a pleasure to thank the ADS team for their always fast and friendly service and inspiring exchange of ideas. My sincere thanks also go to Christopher Erdmann, ESO Library, for reviewing the manuscript and helpful discussions.

References

Accomazzi, A., *et al.*, 2007, Creation and use of citations in the ADS, in: *"Library and Information Services in Astronomy V"*, ed. S. Ricketts, C. Birdie, E. Isaksson, San Francisco, CA, Astronomical Society of the Pacific, ASP Conf. Ser., 377, 93 [arXiv:cs/0610011v1] [cs.DL], http://arxiv.org/pdf/cs/0610011v1

Boyce, P.B., 1998, Electronic publishing in astronomy, in: *"The impact of electronic publishing on the academic community"*, ed. I. Butterworth (London, Portland Press) www.portlandpress.com/pp/books/online/tiepac/session1/ch3.htm

Corbin, B.G., & Grothkopf, U., 2006, LISA – the Library and Information Services in Astronomy conferences, in: *"Organizations and Strategies in Astronomy"*, vol. 7, ed. A. Heck (Kluwer Academic Publishers, Dordrecht), 285 http://www.eso.org/libraries/articles/lisaconferences.pdf

Crabtree, D.R., 2008, Scientific productivity and impact of large telescopes, in: *"Observatory operations: strategies, processes, and systems II"*, ed. R.J. Brissenden & D.R. Silva, SPIE Conf. Proc. 7016, pp. 70161A-70161A-10

Dietrich, J.P., 2008a, The importance of being first: position dependent citation rates on [arXiv:astro-ph], PASP, 120, 224
http://www.journals.uchicago.edu/doi/pdf/10.1086/527522

Dietrich, J.P., 2008b, Disentangling visibility and self-promotion bias in the [arXiv:astro-ph] positional citation effect, PASP, 120, 801
http://www.journals.uchicago.edu/doi/pdf/10.1086/589836

Grothkopf, U., 2003, From books to bytes: changes in the ESO Libraries over the past decade, The ESO Messenger, No. 112 (June 2003), 51
http://www.eso.org/sci/libraries/articles/books2bytes.pdf

Grothkopf, U., & Erdmann, C., 2008, Open access – state of the art, IAU Information Bulletin, 102, 64, http://www.eso.org/libraries/IAU-WGLib/IAU_IB102p64.pdf

Harzing, A.W.K., & van der Wal, R., 2008, Google Scholar: the democratization of citation analysis? Ethics in Science and Environmental Politics, 8, 62
http://www.harzing.com/download/gsdemo.pdf

Hirsch, J.E., 2005, An index to quantify an individual's scientific research output, Proc. Natl. Acad. Sci. USA, 102, 16, 569 [arXiv:physics/0508025v5] [physics.soc-ph]
http://arxiv.org/pdf/physics/0508025v5

Jasco, P., 2005, As we may search – comparison of major features of the *Web of Science*, *Scopus*, and *Google Scholar* citation-based and citation-enhanced databases, Current Science, 89, 1537, http://www.ias.ac.in/currsci/nov102005/1537.pdf

Kurtz, M.J., Eichhorn, G., Accomazzi, A., *et al.*, 2003, The NASA Astrophysics Data System: obsolescence of reads and cites, in: *"Library and Information Services in Astronomy IV (LISA IV)"*, ed. B.G. Corbin, E.P. Bryson & M. Wolf, U.S. Naval Observatory, Washington, DC, 223, http://www.eso.org/sci/libraries/lisa4/Kurtz.pdf

Kurtz, M.J., & Henneken, E.A., 2007, Open access does not increase citations for research articles from The Astrophysical Journal [arXiv:0709.0896v1] [cs.DL]
http://arxiv.org/pdf/0709.0896v1

Mahoney, T.J., 2007, Correspondence to the Editors of "The Observatory", Obs., 127, 401, reprinted: IAU Information Bulletin, 2008, 102, 70
http://www.eso.org/sci/libraries/IAU-WGLib/IAU_IB102p70.pdf

Meho, L., 2007, The rise and rise of citation analysis, Physics World, January 2007, 32
http://eprints.rclis.org/archive/00008340/01/PhysicsWorld.pdf

Meho, L., & Yang, K., 2007, A new era in citation and bibliometric analyses: Web of Science, Scopus, and Google Scholar, subm. to Journal of the American Society for Information Science and Technology, 58 [arXiv:cs/0612132v1] [cs.DL]
http://arxiv.org/pdf/cs/0612132v1

Meylan, G., Madrid, J.P., & Macchetto, D., 2004, HST science metrics, PASP, 116, 790,
http://www.journals.uchicago.edu/doi/pdf/10.1086/423227

Sandqvist, A., 2004, The A&A experience with impact factors, in: *"Organizations and Strategies in Astronomy"*, vol. 5. ed. A. Heck (Kluwer Academic Publishers, Dordrecht), 197. Preprint at http://www.astro.su.se/~aage/OSA5.pdf

Schwarz, G.J., & Kennicutt, R.C., 2004, Demographic and citation trends in astrophysical journal papers and preprints, BAAS, 36, 1654 [arXiv:astro-ph/0411275v1]
http://arxiv.org/pdf/astro-ph/0411275v1

Suber, P., 2007, Open access overview: focusing on open access to peer-reviewed research articles and their preprints,
http://www.earlham.edu/~peters/fos/overview.htm

Viele, P., 2008, Information fluency and physics curriculum: faculty/librarian collaboration, APS Forum on Education Newsletter, Summer 2008
http://www.aps.org/units/fed/newsletters/summer2008/viele.cfm

FROM YOUR PAPER TO VIZIER AND SIMBAD

Laurent Cambrésy[1], Françoise Genova[1], Marc Wenger[1], Cécile Loup[1], François Ochsenbein[1] and Thomas Boch[1]

Abstract. Scientific results strongly rely on previous studies, experiments, and observations. A huge quantity of data is produced by astronomers and needs to be available to the whole community. To support astronomers in their daily research work, the *Centre de Données astronomiques de Strasbourg* (Strasbourg astronomical data center – CDS) collects, verifies, homogenizes, and organizes information – in particular, those published in academic journals – in the most appropriate and comprehensible way. This CDS goal can reach the highest level of quality only through a close collaboration with journals, authors, and referees. This paper presents the succession of processes leading published data to the CDS databases, focusing on the strategy that maintains a high level of quality. Authors and referees are strongly encouraged to actively contribute to this endeavor. A few examples of how CDS databases are used through CDS services are also presented.

1 Introduction

The *Centre de Données astronomiques de Strasbourg* (Strasbourg astronomical data center – CDS) was created in 1972 by the French agency in charge of ground-based astronomy, which is now called INSU (*Institut National des Sciences de l'Univers*), as a joint venture with Strasbourg Louis Pasteur University. Louis Pasteur University merged on January 1, 2009, with the two other local universities to form the *Université de Strasbourg*. The CDS was created as the *Centre de Données Stellaires* (Stellar Data Center), but in 1983 it was decided to extend its scope and to consider all astronomical objects beyond the Solar System.

The high-level goal of CDS is to *"collect, homogenize, preserve and distribute astronomical information for the usage of the whole astronomy community"*. Its initial charter is still relevant:

- collect "useful" data on astronomical objects in electronic form

---

[1] Observatoire Astronomique de Strasbourg, France

- improve them by critical evaluation and combination

- distribute the results to the international community

- conduct research using these data.

From the beginning, CDS has gathered two main types of information:

- astronomical catalogs, with a catalog service shared with several other data centers around the world, which was extended to include tables published in journals. These catalogs were distributed on demand on magnetic tapes and then through an ftp service. The VizieR service, available since 1996, has added a browsing capability (Ochsenbein *et al.* 2000).

- object cross-identification and bibliography were first made available through two different data bases, the *Catalogue of Stellar Identifications* (CSI – Jung & Bischoff 1971; Jung & Ochsenbein 1971), which aimed at cross-identifying a few fundamental star catalogs, and the *Bibliographic Star Index* (BSI – Cayrel *et al.* 1974). SIMBAD (Set of Identifications, Measurements, and Bibliography for Astronomical Data, Wenger *et al.* 2000) was created in 1981 from the merging of the BSI and of the CSI.

The *Dictionary of Nomenclature of Celestial Bodies outside the Solar System* (Lortet *et al.* 1994) and the Aladin interactive star atlas (Bonnarel *et al.* 2000) are two other important CDS products which will be referred to in the following.

Involvement of active researchers in the CDS staff has always been recognized as an important factor for success, since it is the only way to ensure the required scientific expertise on the service content and to make sure that the services are suited to the constantly evolving needs of the scientific community. To maintain a high level of scientific activities, the CDS was created inside and as a part of an Observatory, which provides an active research environment. The CDS team integrates astronomers, computer engineers who manage the databases and the user interfaces, and highly specialized bibliographers, who are called *documentalistes* in French – this shows that the CDS librarians are specialized in extracting "information" from and building metadata for documents. Information and metadata of interest for CDS are not only the usual bibliographic information managed in libraries (authors, publication references, etc.), which constitute the relatively easy part of the work. The point here, as will be shown in the following, is to check the description of the catalogs and tables stored in VizieR or to extract important data about astronomical objects for SIMBAD.

In the following, CDS services and procedures are described with the aim of showing how they are built and how the scientific community can help the data center in its constant quest for quality, along with a few hints to how they can be used. Only a selected set of information about the CDS services is highlighted in this paper. For more details, the reader is referred to Genova *et al.* (2000) and to the other papers of the series published in the April 2000 Special Issue of *Astronomy & Astrophysics Supplement Series* (Vol. 143, No. 1, several of which have been cited above): "The CDS and NASA ADS resources: New tools for astronomical research", and for a global more recent view of CDS history and evolution, to Genova (2007).

2   The VizieR Database

2.1   Content

The link between VizieR and SIMBAD is not always clear to new users since the two databases are indeed very different. SIMBAD results from the cross-identification of major catalogs, tables and individual objects published in the literature. VizieR comes from a different approach: the astronomical catalogs are kept in their original form, in what we call the *FTP archive* (a set of files and directories openly accessible from cdsarc.u-strasbg.fr). In this repository, one catalog occupies one directory and is always accompanied by a *standard description*, in the form of a file named ReadMe described in the next Section 2.2. These homogeneous descriptions are the key to the VizieR system, a relational database gathering the tabular data (catalogs) existing in the FTP archive; in addition, VizieR gives access to very large surveys that are not part of the FTP archive because of their huge size.

VizieR currently provides access to the most complete library of published astronomical catalogs and data tables available online (more than 8000 catalogs in June 2010), organized in a self-documented database. A brief summary of VizieR is given below (Sect. 2.3); the contents of VizieR are illustrated by Figure 1, which represents the spatial distribution of $\sim 10^9$ sources available in VizieR after removal of the all-sky surveys.

2.2   The ReadMe File

Information that describes the data – the metadata – is traditionally presented in the introduction of the printed catalog, or explained in one or several published papers presenting and/or analyzing the cataloged data. Metadata play a fundamental role. First they provide scientists with information about the context of the data so that they can make their judgment about the suitability for their project. Also a minimal knowledge of the metadata is required by data processing systems in order to merge or compare data from different origins – for instance, the comparison of data expressed in different units requires a unit-to-unit conversion, which can be performed automatically only if the units are specified unambiguously.

This need for a description that is readable both by a computer and by a scientist led to a standardized way of documenting astronomical catalogs and tables, promoted by the CDS and adopted by other data centers in the form of a dedicated ReadMe file associated to each catalog (Ochsenbein 1994). The ReadMe description file (see example Fig. 2) starts with a header specifying the basic references of the catalog – title, authors, references – and contains a few key sections introduced by standard titles like *Description* or *Byte-by-byte Description*

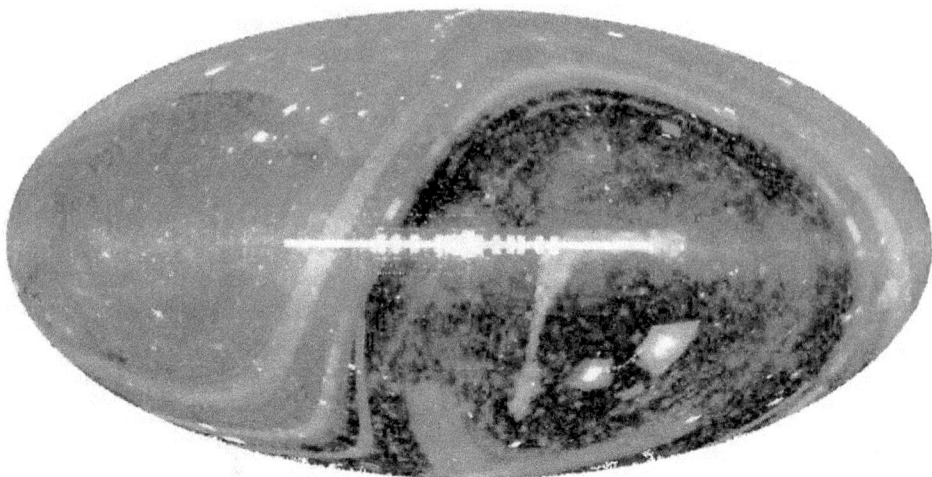

Fig. 1. VizieR surface density map in galactic coordinates, representing $\sim 10^9$ sources after removal of all-sky surveys USNO-B1 ($\sim 10^9$ sources) or 2MASS. The horizontal bar in the central region is due to the GLIMPSE catalog (map built on June 2010 with more than 7500 catalogs).

of file. Such a file is relatively easy to produce by someone who knows the catalog contents – or better by the authors themselves.

The most important part of the ReadMe file is the *Byte-by-byte Description*, which details the table structures in terms of formats, units, column naming, or labels (some conventions are used for label assignments for consistency), existence of data (possibility of unspecified or null values), and brief explanations.

This standardized way of presenting the metadata proved to be extremely useful, especially for data-checking and format conversion. Many errors were detected in old catalogs simply because a general checking mechanism became available.

Astronomy & Astrophysics authors are expected to supply the documentation of their data in this simple form. Template files, as well as a few tips on how to create the ReadMe file, are accessible on the VizieR webpage[1]. The ReadMe files and the data files are then checked by a specialist who contacts the authors if errors are detected or when changes are necessary to increase the clarity or homogeneity of the description.

The data accessible in the *FTP Archive* are not restricted to tabular data: if historically astronomical catalogs have been the first type of data distributed on electronic media by the CDS in the early seventies, the astronomical data available from this repository now include also spectra, images, data cubes, etc. However, data on the public repositories are in principle restricted to *scientific data, i.e.,* data that can be re-used in the data processing tools used in astronomy. This

---

[1]http://vizier.u-strasbg.fr/doc/submit.htx

```
J/A+A/466/137        JHKs photometry in Westerlund 2        (Ascenso+, 2007)
================================================================================
Near-infrared imaging of Galactic massive clusters: Westerlund 2.
    Ascenso J., Alves J., Beletsky Y., Lago M.T.V.T.
    <Astron. Astrophys. 466, 137 (2007)>
    =2007A&A...466..137A
================================================================================
ADC_Keywords: H II regions ; Clusters, open ; Photometry, infrared
Keywords: open clusters and associations: individual: Westerlund 2 -
          stars: pre-main sequence - infrared: stars

Description:
    Photometry for the stars detected with SOFI, NTT in J, H and Ks in the
    field containing the massive cluster Westerlund 2 (also know as RCW 49)
    associated with the HII region Gum29. For each star detected in all
    three bands the table contains the ID, equatorial coordinates, J, H
    and Ks magnitude and photometric errors. For the stars detected only
    in Ks or in H and Ks the magnitude and photometric error fields in the
    absent bands are filled with the values 99.99 and 9.999 respectively.

File Summary:
--------------------------------------------------------------------------------
 FileName   Lrecl  Records   Explanations
--------------------------------------------------------------------------------
ReadMe        80       .     This file
w2phot.dat    65     5727    Positions and magnitudes of stars in Westerlund 2
--------------------------------------------------------------------------------

Byte-by-byte Description of file: w2phot.dat
--------------------------------------------------------------------------------
   Bytes Format Units   Label     Explanations
--------------------------------------------------------------------------------
   1-  4  I4    ---      Seq      Identification number
   6-  7  I2    h        RAh      Right ascension (J2000.0)
   9- 10  I2    min      RAm      Right ascension (J2000.0)
  12- 16  F5.2  s        RAs      Right ascension (J2000.0)
      18  A1    ---      DE-      Declination sign (J2000.0)
  19- 20  I2    deg      DEd      Declination (J2000.0)
  22- 23  I2    arcmin   DEm      Declination (J2000.0)
  25- 29  F5.2  arcsec   DEs      Declination (J2000.0)
  31- 35  F5.2  mag      Jmag     ?=99.99 Magnitude in the J band (1)
  37- 41  F5.3  mag      e_Jmag   ?=9.999 Photometric error in the J band (1)
  43- 47  F5.2  mag      Hmag     ?=99.99 Magnitude in the H band (1)
  49- 53  F5.3  mag      e_Hmag   ?=9.999 Photometric error in the h band (1)
  55- 59  F5.2  mag      Ksmag    ?=99.99 Magnitude in the Ks band (1)
  61- 65  F5.3  mag      e_Ksmag  ?=9.999 Photometric error in the Ks band (1)
--------------------------------------------------------------------------------
Note (1): The values 99.99 for the magnitude and 9.999 for the errors
     correspond to sources for which we do not have photometry in the
     corresponding band.
--------------------------------------------------------------------------------

Acknowledgments:
    Joana Ascenso, joanasba(at)astro.up.pt
================================================================================
(End)       Joana Ascenso [CAUP, Portugal], Patricia Vannier [CDS]   25-Jan-2007
```

Fig. 2. Example of ReadMe file for catalog I/221.

restriction means that FITS images, for instance, can be found (FITS includes essential details about origins, sky location, etc.), but JPEG or PDF images are generally excluded.

2.3   VizieR

Once a catalog, as described according to the above conventions, is part of the *FTP archive*, it becomes visible to any interested scientist. Search engines provided by the CDS help to locate these catalogs from their textual contents with *GOOGLE*-like queries. Textual searches, however, cannot answer to questions like *"find in all tables of this archive which contains entries related to point sources in a 2-arcmin region around my target"*. Such queries have to deal with numeric values and have to know what a position is and which coordinate system is used in each table, etc.

These queries are possible with VizieR, the database created from the contents of the *FTP archive*. VizieR is basically a relational database that keeps both the original data (each catalog or table of the FTP archive becomes a table of the database) and the associated metadata extracted from the ReadMe files (stored in a set of dedicated tables). These metadata play the essential role of giving a meaning to all the numbers and symbols included in the catalogs and tables. VizieR structure permits queries to be constrained by any parameter, *i.e.* column, of the tables. Some further metadata homogenization is performed to better interpret the parameters contained in the thousands of tables of VizieR, essentially by assigning a *Unified Content Descriptor* (UCD) to each column of each table. The UCD, originally defined at the CDS (Ortiz *et al.* 1999) is shorthand to qualify the meaning of a parameter in a thesaurus of the existing parameters used in astronomy (*e.g.* positions, magnitudes, oscillator strength), which is now developed in the framework of the *Virtual Observatory*.

Besides the homogenized metadata, VizieR also introduces *links* to associated data. As an example, when a paper contains images or spectra associated to observed or modeled objects, VizieR may provide a link to the images or spectra related to the corresponding object entry in the table. Such links may also point to other VizieR tables, to other databases (SIMBAD, NED, HyperLeda), or to repositories (SDSS, HST archive, etc.).

VizieR also gives access to large surveys, such as 2MASS or USNO-B1. The access to these huge surveys, which is not possible in the *FTP Archive* because of the volume of data, is optimized to give fast answers to positional queries.

Currently (June 2010), VizieR contains more than 8000 catalogs made of 19 000 different tables for a cumulative number of 3×10^5 columns, *i.e.*, an average of 2.3 tables per catalog and 12.5 columns per table. VizieR responds to an average of 1.2×10^5 queries/day.

2.4   Usage of VizieR

Access to VizieR by Web is available by script, or by programs. The data published in VizieR also become accessible by the Virtual Observatory, which means that the tabular material, once properly documented and ingested into VizieR, becomes directly accessible by any tool aware of these standards. As a practical example, tables describing astronomical sources become visible in Aladin (Sect. 5). The results available this way are also easily retrieved for the creation of large samples required to derive new astronomical results.

An accurate description of the data related to astronomical objects is also an important step before integrating them into SIMBAD (see Sect. 3.3); therefore, do not miss the documentation of your data!

3 How your Data are Integrated in SIMBAD

3.1 Acronym Creation in SIMBAD

Astronomical object designation, either called name or identifier, is a key point for SIMBAD since it refers unambiguously to an object. Some rules need to be defined for clarity. The general syntax of an identifier is the abbreviated catalog name, or acronym, followed by one or several fields, which can be numbers, coordinates, strings, etc. An identifier is case and space insensitive.

When no general catalog name can be used, the acronym is directly based on the author's initials between square brackets with the year of publication of the paper. For example, HD 97300 is also named [FL2004] 44 after Feigelson & Lawson (2004).

Object names, such as Vega and Altair, but also Barnard's star, Crab Nebula, Sgr A, HDFN, or HDFS, are stored in the database in a specific catalog called "NAME", while star names in constellations, such as Lyra, are stored in the catalog "*" (*e.g.* * gam Lyr), and variable stars (such as RR Lyrae) in the catalog "var" (also called "V*"). To some extent, identifiers can be shortened (*i.e.* NAME can be omitted), or can be typed according to some accepted usage. A set of rules, based on regular expressions and substitutions, try to normalize the typed identifier before querying the database.

Greek letters should be abbreviated as three letters: alf, bet, for α and β, but also mu., nu., and pi. (with a dot), for μ, ν, and π. This is required to avoid ambiguities in some identifiers. Constellation names should be abbreviated with the usual three letters: alf Boo, del Sct, FG Sge, NOVA Her 1991. The full list is available online.

Identifiers of a multiple system may generate a list of all objects belonging to the system. For instance, ADS 5423 calls for the four components, A to D, of the stellar system around Sirius. This is only true for some specific identifiers.

Clusters that have no NGC or IC number are named under the generic appellation Cl followed by the cluster name and number: *e.g.*, Cl Blanco 1 is the 1st stellar cluster named by Blanco. Stars in clusters follow complex *historical* rules. They may belong to a *main* designation list or to subsequent lists. NGC 5272 692 is star 692 in the list by Von Zeipel, considered as the main list. Subsequent lists have designations starting with Cl* such as Cl* NGC 5272 AC 968 (list by Aurière & Cordoni), Cl* Melotte 25 VA 13 (13th star in the list by van Altena for Melotte 25 – the Hyades cluster).

3.2   Integration of Objects Cited in the Text of Papers into SIMBAD

As a general rule, all objects cited in a paper are stored in SIMBAD with a reference to this paper and relevant data if any. This has long been done by *documentalistes* reading journals and checking them for object citations. Additional paths for object entry have been made possible by the development of electronic journals. In particular, the CDS and *Astronomy & Astrophysics* collaborate to provide links between object names in papers and SIMBAD, by asking authors to tag *important* object names. These tags are used to identify objects to be referred to in SIMBAD (Sect. 3.2.1). Most of these objects, however, are entered through a systematic scan of the papers or by a semi-automated processing of published tables. The procedure used for naming objects and cross-identifying them is described in Sections 3.1 and 3.2.3.

3.2.1   Tagging Astronomical Objects

The *Astronomy & Astrophysics* instructions ask authors to *"surround any astronomical object in your text, as well as in short tables with the command"*:

```
\object{<objectname>}
```

Objects tagged with this macro will appear linked to the corresponding SIMBAD page in the journal electronic edition. Only the authors are allowed to tag the objects. About 20% of the references in *Astronomy & Astrophysics* contain tagged objects, which explains why many astronomical objects do not have a SIMBAD link in the electronic edition.

    The list of publications that cite an object is provided for each entry in SIMBAD. Although the tag does not require having a paper included in the list of references associated with an object, it significantly shortens the delay of this process. Tagging is therefore suitable for rapidly informing the community interested in an object that a specific paper is relevant.

    Tags seem an efficient tool for allowing authors to directly interact with the SIMBAD content, however reality is not that simple. Naming an object is indeed a critical step that can often lead to ambiguities. Each tag requires a check by the SIMBAD team at the time of the paper's electronic publication, because it can be a new name, which is not in SIMBAD yet, or because of a potential careless use of nomenclature by the author **and** a lack of attention from the referee. Regarding the authors' part, CDS provides a web service to perform all the verifications by submitting the *.tex or the *.obj files (the .obj is generated by the compilation of the LaTeX file):

```
http://vizier.u-strasbg.fr/viz-bin/Object
http://vizier.u-strasbg.fr/viz-bin/Sesame.
```

The *Astronomy & Astrophysics* instructions to authors provide these two addresses and *"encourage you to test the astronomical objects using the sites and easy tools available at the CDS"*. Journals from the *American Astronomical Society* have

object	OType	J2000 position	Nrefs	Resolver
perseus	Assoc*	02:21:00.00 +57:36:00.0	77	Simbad: Ass Per OB 1-
perseus molecular cloud	MolCld	03:36:00.00 +31:00:00.0	101	Simbad: NAME PER CLOUD
perseus cluster	ClG	03:19:48.00 +41:30:42.0	791	Simbad: ACO 426

Fig. 3. Summary of objects tagged in a test manuscript with the \object macro. This is obtained with the tool available at the URL http://vizier.u-strasbg.fr/viz-bin/Object.

adopted similar methods for maintaining links from the electronic publications to SIMBAD. Despite these tools, the number of necessary actions to build a working link is increasing. In 2002, 13% of the papers had tags and 25% of them needed *corrections*. In 2006, 25% of the papers had tags and 45% of them needed *corrections*. *Corrections* mean incorrect designations or creation of new acronyms for SIMBAD (new objects). The latter case is obviously not an author's mistake. The increase in corrections is a consequence of tagging the objects present in long tables. Since these objects, only cited in long tables and not explained in the text, are not individually important for the paper, authors are now asked to tag only the most relevant objects, meaning no more than 10–20 tags per paper. Long tables must be excluded systematically. In the present paper, tags are not used at all because none of the cited objects are studied, and this reference is obviously not relevant to any science topic.

Besides this, correct designations are sometimes ambiguous because they may refer to several different possible objects, for instance, the result of a SIMBAD query for *Perseus* yields 17 possible entries, including a molecular cloud, a star association, a superbubble, a spiral arm, a Seyfert 2 galaxy, or even a galaxy cluster. It is therefore critical to avoid ambiguous names. As shown in Figure 3, *Perseus* alone will point the Perseus OB 1 star association, so more specific names must be chosen to reach the correct target, hence the importance of checking links before submitting the paper.

3.2.2   Designations of Astronomical Objects

SIMBAD is not limited to tagged objects. All papers from ∼30 journals are read by the CDS staff to extract all astronomical objects cited in the text. Taking the example of *Perseus* again, staff have to decide which of the possible objects the paper actually deals with. The first step is to recognize an astronomical name in the text, given the prolific imagination of some authors. One might naively think that any word with special characters or numbers is the signature of an astronomical object, but this is not necessarily true. A field designation for instance will probably have the typical structure of an object name although its inclusion in SIMBAD is not desired. Also some object type names are based on their prototype, such as T Tauri stars or FU Orionis stars. When the star name is used as object type, the star T Tauri or FU Ori is not to be added to the list of objects for the reference, and, symmetrically, the reference should not be added to the star T Tauri or FU Ori. The context of the paper is usually clear enough on this aspect for a well-trained *documentaliste* to find out.

Confusing names such as LW1 are more difficult to track. LW1 is an ISOCAM filter onboard the ISO satellite, and it is also a globular cluster named NGC 1466. LW1 stands for Lynga & Westerlund (1963) source number 1 and comes from a table of 483 sources with a record number as a first column. If this paper were being ingested at the CDS now, the LW1 name would be [LW63] 1. Using a prefix formed by the author's initials with the year of publication within square brackets has the advantage of linking the designation to a publication and prevents any confusion.

For example, if one decides to name his/her object M1, SIMBAD will remove the confusion with the famous Messier 1, the Crab nebula, by using the prefix. Unfortunately, some authors might cite the object but drop the prefix, and it will become more and more difficult to know which M1 it is, except maybe for the specialists of the relevant topic. Confusion will spread over the literature. Another example presented in Figure 4 shows the content of the dictionary of nomenclature[2] for the identifier L. The second column gives the actual designation used in SIMBAD. Lynds dark nebulae should be called LDN followed by a number, such as LDN 113, for instance. The name encountered is often L113 instead. This happens so often that the ambiguity is actually solved by SIMBAD where L113 is correctly interpreted as LDN 113. Nevertheless, if one writes L113, thinking to the open cluster Lindsay 113, the result will be wrong.

On the other hand, the square-bracket prefix is not suitable for *official* catalogs. It would be meaningless to add a prefix to SDSS or 2MASS sources since the designation is formatted, so unique, and comes from a released catalog. There cannot be any confusion. Actually there are, because some authors decide to alter the official name by truncating the correct designation or by changing one digit (when the format is based on celestial coordinates). This should be absolutely forbidden. Official names must be respected even if based on coordinates, regardless of the slightly different coordinates found by follow-up observations.

Finally, papers would gain in clarity if astronomical objects were systematically accompanied by coordinates or with a second known designation in case of cross-matches. This simple recommendation would avoid a lot of ambiguities in the object identification.

3.2.3   Cross-Identifications Proposed in Papers

Among the recent progress in astronomy, multi-wavelength analysis is certainly one of the most spectacular. This approach has been supported by SIMBAD since its beginning in the parent database CSI. One of the main SIMBAD goals is to

---

[2]Dictionary link on `http://cdsweb.u-strasbg.fr/`

Acronym	Use Format	Year	1st Author	Obj. Type
L	L FFFF-NNNNA	1942	LUYTEN W.J.	PM*
(L)	[AEL81] {L}NNa	1981	ALLOIN D.+	Part of G
(L)	[H68] {L}	1991	WAKKER B.P.+	HVC
(L)	[L63] NNN	1963	LINDSAY E.M.	Em*
(L)	[L65] N	1965	LYNGA G.	HII
(L)	[L92] HHMMSS.s+DDMMSS {Sgr D} N {Sgr E} NN	1992	LISZT H.S.	(Rad)
(L)	[L92b] {SA} FFF LN {TPhe} A {Mark} AN {RU} NNN A {PG} HHMM+DDd A	1992	LANDOLT A.U.	*
(L)	Lal NNNNN	1950	BAILY F.	*
(L)	LAWD NN	1949	LUYTEN W.J.	WD
(L)	LBN LLL.ll+BB.bb	1965	LYNDS B.T.	HII
(L)	LDN NNNNA NNNNWW	1962	LYNDS B.T.	Dark N
(L)	[LGC91] {ACO} NNNN NNNN	1991	LUCEY J.R.+	G + * in ClG
(L)	Lin NNNa	1961	LINDSAY E.M.	Em. Obj.
(L)	Lindsay NNN	1958	LINDSAY E.M.	OCl
(L)	LLNS NNNNN	1976	LODEN L.+	*
(L)	Loden NNNN	1977	LODEN L.O.	Loose and,or poor OCl
(L)	[PWA92] NN	1992	PRUSTI T.+	*
NLTT	L FFFF-NNNNA	-	LUYTEN W.J.	PM*
OL	Ohio NNN.N -NNN.N	1967	SCHEER D.J.+	(Rad)

Note: Usage of acronyms in parentheses like 'L' should be *avoided*

Fig. 4. Dictionary of nomenclature for the acronym L.

ingest the cross-identifications found in the literature. The scientific impact is considerable for characterizing of astronomical objects and for understanding the underlying physics. Cross-identifications will eventually lead to spectral energy distributions. Because of their crucial role, cross-identifications are introduced as carefully as possible in SIMBAD.

Unfortunately, the literature cannot always be trusted. A nearest neighbor may be a reliable cross-identification if wavelengths, sensitivities, and resolutions are similar. If not, this technique is not sufficient so a scientific analysis is required. For instance, the cross-identification of a 2MASS source (arcsec resolution) with an IRAS source (arcmin resolution) can be very hazardous, except if the object type, say a carbon star, ensures that a single 2MASS object can explain the far-infrared flux measured by IRAS. Also the cross-identification of optical with radio or X-ray sources can be risky if the method is limited to the nearest-neighbor criterion. Since authors are specialists in their domains, we cannot reach the equivalent level of expertise in all fields of astronomy to decide whether a cross-identification proposed in a paper can be trusted, but we try to ignore hazardous matches as much as possible.

Incomplete cross-identifications are often found in papers. A typical author's choice, for instance, is to provide a list of sources in the optical and to add the SDSS magnitudes without providing the SDSS names. There are definitely cross-matches, but they are implicit and we might have trouble finding out why.

Ambiguities (*i.e.* two nearby objects) generally occur, implying a serious loss, sometimes waste, of time for the SIMBAD team when a lot a papers are waiting to be ingested, and loss of useful information known to the authors.

3.3 Integration of Catalogs and Tables in SIMBAD

While short published tables are integrated in SIMBAD manually by the bibliographers in charge of those articles, longer tables listing more than about 50 objects are integrated in a semi-automated way by other specialized bibliographers. Such tables are first archived electronically in the VizieR database. These VizieR electronic tables are then used and processed to include the objects and new data in SIMBAD in the most consistent way. In the first step, cross-identifications with objects already in SIMBAD are systematically searched, using coordinates and names, and checking the consistency between object types. Most of the work of the bibliographers consists in adjusting the search parameters to optimize the cross-identifications, and in checking ambiguous cases individually, with the support of Aladin facilities and of the expertise of CDS astronomers. Once all problems have been solved, the integration of the tables in SIMBAD is done automatically. Although such a procedure can be time consuming, it is one the major goals of SIMBAD to perform careful cross-identifications between various catalogs. In case of doubt, the conservative rule is to create a new object rather than to risk an erroneous cross-identification.

In past years the number of tables published in the articles has dramatically and continuously increased. Among all the tables archived in VizieR, about one third does not have to be integrated into SIMBAD, because they are not relevant (*e.g.* list of atomic transition probabilities) or because it would not improve or would even degrade the level of information: lists of detailed measurements, lines or variability studies for instance; photometric CCD surveys without characterization of the sources in very crowded fields (like in globular clusters); extremely huge catalogs like 2MASS or the SDSS. Pieces of huge catalogs are integrated via shorter tables of follow-up studies, or sometimes by dedicated operations, to cross-identify the content of SIMBAD with new large catalogs. Still, since 2005 the CDS has faced a total of 400–450 tables to be integrated into SIMBAD, with sizes ranging from a few 10 to 10^5 objects.

The global CDS strategy is to take full advantage of the complementarity between VizieR and SIMBAD: all tables are made available in VizieR, and a large fraction of them in SIMBAD – in priority those for which the SIMBAD added value is the most prominent.

To avoid too much delay in the integration of more important tables or catalogs, the strategy is that the CDS team, astronomers and bibliographers, perform a scientific expertise of all tables. The fundamental data and measurements to be entered in SIMBAD, as well as the main parameters to search for cross-identifications, are setup by the team, as is an order or priority. Then tables are classified into 3 groups: group 1 (first priority, 45%), to be integrated in SIMBAD as soon as possible; group 2 (second priority, 45%), to be entered in SIMBAD when possible; group 3 (10%), lists of known objects without new data, which can be done entirely automatically. Various criteria can yield a first-rank priority. In particular, the level of information input, which includes accurate redshift or velocity measurements, memberships, expert cross-identifications especially between different wavelength ranges, and improvement of data (coordinates, spectral types, magnitudes, etc.). The priority also goes to outstanding fields of research, such as cosmological deep fields, brown dwarfs, X- or gamma-ray researches, and identification of YSOs. The number of citations (relative to the date and journal of publication) and number of catalog queries in VizieR are checked as well.

The content of the published tables is fundamental to gaining or wasting time and information. For instance, a table listing SDSS sources can be integrated into SIMBAD two to three times faster if the authors use the latest SDSS version and list the correct and entire name. Careful cross-identifications, expertized by the authors, are also very valuable, because they save time by including the table and improve the level of information. The content of the SIMBAD database is made not only by the CDS team, but also by the astronomers themselves through the information included in their papers and the tables therein.

Information from tables is thus stored in VizieR and/or SIMBAD. To help users retrieve the information available at the CDS more easily, the team is now working on a general portal able to access all CDS resources, including of course VizieR and SIMBAD, from a single query.

3.4   Hierarchical Links

In SIMBAD, some tentative hierarchical links between astronomical objects were first introduced through object types, such as a star or galaxy in a cluster or group. This, however, has turned out to be inadequate, since it did not allow all cases of hierarchy to be included and was often reflecting more an observing method than a real link. In particular, CCDs surveys of clusters easily lead to a field of *stars in cluster* distributed according to the instrument field of view rather than the actual cluster shape. Finally, the link by object type did not allow keeping track of either the reference in SIMBAD or the parent body, if relevant. A new tool was thus developed to include hierarchical links between SIMBAD objects. The link provides the following information: status (*parent* or *child*), probability of membership (given in the reference or estimated by SIMBAD), and bibcode of the reference providing the link. Hierarchical links can be physical, like objects belonging either to a cluster or a galaxy, multiple systems, or to a shell and or central star of a supernova. They can also be non-physical due to the improvement of instruments and resolution. For instance, an IRAS source can be resolved in several components in MSX; then, the IRAS source is the *parent* of the MSX sources, although it is no longer a single astronomical object but historic artifact worth keeping track of.

Thus, when a link is indicated on a SIMBAD object, it does not mean that the two objects are actually physically linked: a link is also put for foreground or background sources with a low or no membership in order to save this information of non-membership which can be as important as membership for some studies. In some cases, one astronomical object may have several parents. For instance, one candidate Wolf-Rayet (WR) star in a galaxy was later resolved into one B star and one WR star; the parents of both stars are the historical WR candidate and the galaxy. Hierarchical links have been systematically entered in SIMBAD as a test since 2008 for tables and catalogs archived in VizieR (Sect. 3.3). No hierarchical links are put for short lists of objects entered manually by the bibliographers, as it would always require the expertise of an astronomer. We do not consider performing a full update of the SIMBAD database for hierarchical links, which would be incredibly time-consuming. However, users can send a request to CDS[3] proposing to include them for a dedicated catalog or areas, and providing us with as much information as possible.

4 The SIMBAD Database

This section focuses on the user's point of view by describing the content of SIMBAD to illustrate the final result of the data entry process described in the previous sections. The query modes are described in the Appendix.

4.1 Content

SIMBAD is continuously updated and contains 4.8 million objects in June 2010, with several kinds of data.

- **Identifiers**: census of the different names found in the literature to designate an astronomical object. The list relies on successive cross-identifications. The total number of identifiers in SIMBAD reaches 13.9 million.

- **Basic data**: mostly data that allow the identification of the object, *i.e.* coordinates, proper motions, parallaxes, radial velocity and redshift, spectral type, morphological type, galaxy dimensions, and magnitudes.
  The object type is also provided, together with a list of complementary, or historical, object types inferred from the identifier list.
  Objects can have *notes* to inform the user of some mistakes found in the literature, or a link with some other object, or simply more extensive information connected with this particular object.
  Hierarchical links are proposed as links between a *parent* object and its *children*, for instance, a double star and its components, a cluster of galaxies and the galaxies it contains, a star cluster and its stars.
  Similarly, *association links*, still under development, will specify the association of two astronomical objects for any possible reason described in a paper.

---

[3]Mail to: question@simbad.u-strasbg.fr

- **Bibliographic references**: associated through citation in the paper of any of the object's identifiers. The bibliography gives access to abstracts and electronic articles when available either directly from publishers or through ADS[4]. Reciprocally, SIMBAD bibliography is systematically included in ADS.

- **Observational data**: also called *measurements*, they are always associated with a bibliographic reference. New entries have been recently added, namely distances and star diameters.
  Beside, external links are provided for some identifiers. Currently SIMBAD has a link to HEASARC[5] triggered by the existence of an identifier from a High Energy catalog and a link to some catalogs in VizieR (such as IRAS, 2MASS, HD, GSC, Hipparcos) and a link to the NED[6] database.

4.1.1 Identifiers

The identifiers constitute the basis for the SIMBAD name resolver facility that provides, in response to any object name, the coordinates corresponding to the object position or the list of papers citing the object. The name-resolving power of SIMBAD is used by many archives and information systems (such as the archives of Hubble Space Telescope or European Southern Observatory, the High Energy Astrophysics Science Archive Center, the Astrophysics Data System, servers of the Digitized Sky Surveys).

4.1.2 Basic Data

- Object types: from star to maser source, or cluster of galaxies, some 180 different categories, general, or very specific, are proposed. The object type list refers to a hierarchical classification of the objects in SIMBAD derived by the CDS team (Ochsenbein & Dubois 1992). It is evolving by adding new object types when required by the evolution in astronomy. For instance, the development of the ultra luminous X-ray sources (ULX) research domain yielded the creation of a dedicated object type in SIMBAD. A complete list is available online on the SIMBAD website.

  The object types are also a powerful tool for data cross-checking and quality control. They are also used in an experimental project of building an ontology for checking SIMBAD objects consistency in the frame of the Virtual Observatory.

- Coordinates: they are stored in the International Celestial Reference System (ICRS, see Feissel & Mignard 1998) at epoch 2000.0, after the publication of the Hipparcos and Tycho catalogs.

---

[4] Astrophysics Data System: http://ads.harvard.edu
[5] High Energy Astrophysics Science Archive Research: http://heasarc.gsfc.nasa.gov
[6] Nasa/IPAC Extragalactic database: http://nedwww.ipac.caltech.edu/

- Magnitudes: SIMBAD allows storage of any number of magnitudes and fluxes in the Vega or the AB photometric system. Historically, SIMBAD has stored B and V magnitudes. Currently this list has been extended to 13 magnitudes: U, B, V, R, I, J, H, K, and SDSS magnitudes u, g, r, i, z. They are provided with the associated uncertainty and bibliographic reference.

4.1.3 Bibliographical Data

The creation of a list of papers citing an astronomical object requires a code to refer to each publication. A collaboration between SIMBAD and NED has led to the definition of the so-called *bibcode* assigned to every paper. This 19-digit bibcode contains enough information to locate the article (including year of publication, journal, volume, and page number) and has been rapidly adopted by journals and ADS. With the development of electronic publications, the page number is becoming obsolete, so too the bibcode. Publishers have defined a generic code, the digital object identifier (DOI), which is more adapted to current publication processes. SIMBAD will obviously follow this natural evolution.

The access to the list of references for any object is one of SIMBAD major products for users, but for many objects, there are many references where it may be difficult to find those of interest. To increase its usefulness, a flag is being added to represents the relevance of the citation in terms of astronomical contents for each paper linked to an object, so different relevance levels are possible: paper entirely devoted to an the object, object cited in the title, object in the abstract, object cited in a table, or object simply cited in the paper. This new flag is being tested and will likely be added for most new papers.

Several types of comments can be associated with the references in SIMBAD and are displayed after the reference:

- General comments: they are often comments added by the bibliographers about the problems encountered while cross-identifying the objects mentioned in the paper, typos in object names, etc.

- Notes about the existence of associated electronic tables or abstracts in the CDS server. Papers including no object are also flagged.

- Information on how the quoted objects are named in SIMBAD (comments related to the Dictionary of Nomenclature of Celestial Objects).

4.1.4 Some Statistics on the Data Contents

The astronomical content of SIMBAD results from adding a selection into the database of *important* catalogs and published tables and a survey of the complete astronomical literature. This can be illustrated by a few histograms of the V magnitudes (Fig. 5) and of the 13 magnitudes already available in SIMBAD (Fig. 6). The list of the 30 catalogs for which SIMBAD has the most identifiers is displayed in Table 1.

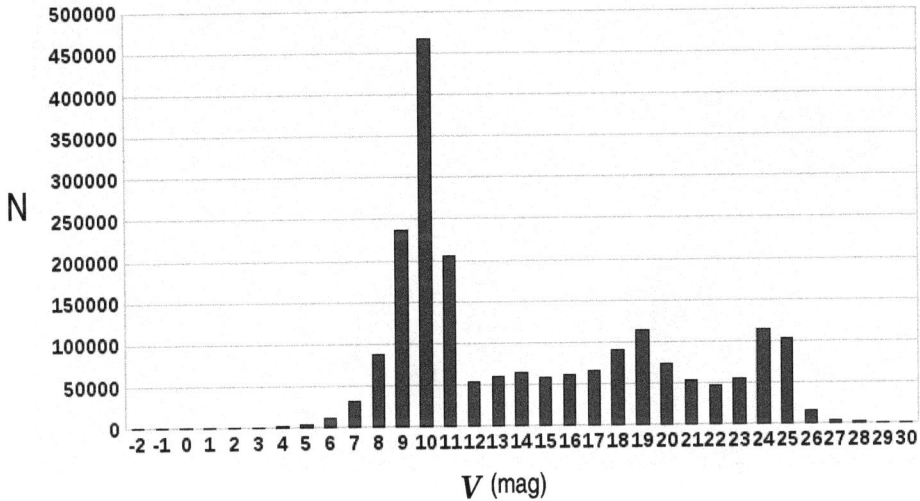

Fig. 5. *V* magnitude histogram. SIMBAD is complete up to $V \approx 10$ mag (since inclusing of the Tycho catalog) and also contains very faint objects from the deepest fields observed so far.

Table 1. Main catalogs in SIMBAD with the corresponding number of identifiers in the database. A single source generally has several identifiers from different catalogs.

2MASS	1372142	FIRST	247749	HIC	118205
GSC	988949	2MASX	240411	HIP	118161
TYC	947069	CPD	231232	LCRS	92991
COSMOS	597301	CD	212186	CPC	86794
PPM	468709	NVSS	197587	[VV2006]	85219
Cl*	381965	LEDA	195290	GEN#	82344
HD	343325	AG	183566	UBV	73067
BD	319291	SDSS	162665	USNO	72798
IRAS	301189	YZ	137843	MGC	63309
SAO	258986	1RXS	124790	LSPM	61928

4.2 SIMBAD Access

SIMBAD query modes are described in detail in the Appendix, to show how the wealth of available information can be accessed by users by simple to advanced queries, depending on their needs. The main access mode is the web, on which each query mode has its own form. Apart from the regular use of filling out a query form and getting the result in a webpage, the user has the possibility of submiting input files and to get the result in an ASCII file for further processing. SIMBAD can be accessed in two different ways: defining an URL with all requested parameters

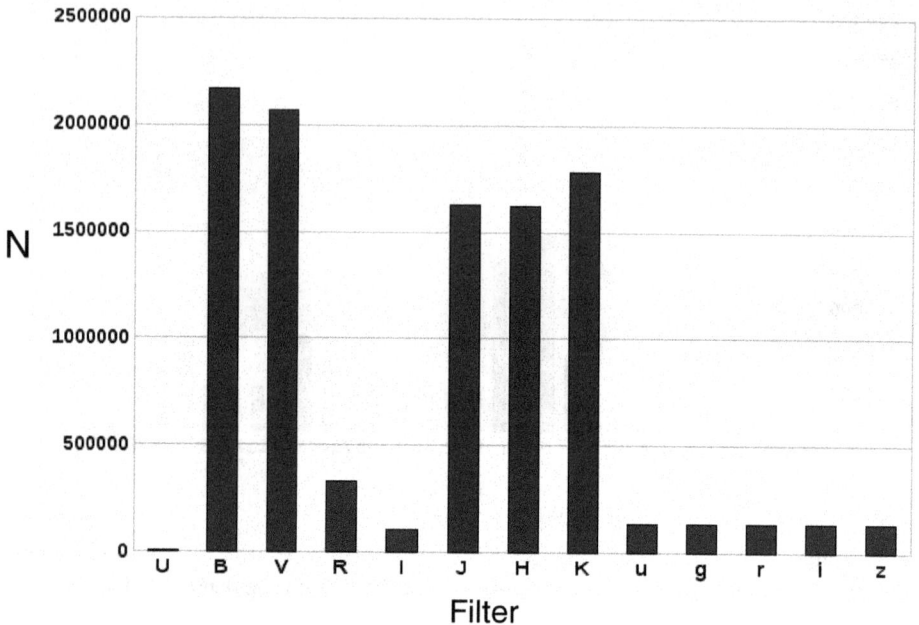

Fig. 6. Histogram of all available magnitudes. Historically only B and V magnitudes have been stored in SIMBAD but this is now extended to more passbands. The high proportion of sources with JHK magnitudes comes from a dedicated operation of massive cross-identification between SIMBAD and 2MASS performed at the CDS, which added the near-infrared magnitudes for $\sim$1.6 million sources.

and getting the output as an HTML page or an ASCII formatted result; or using the SIMBAD defined web services.

5 Aladin

A web query to VizieR or SIMBAD produces a page containing information about an object or a list of objects. An interesting alternative is to investigate sky positions and to display the results of several queries simultaneously, superimposed on images. A few details on Aladin are given here, as an illustration of the possible use of data from VizieR and SIMBAD in a software tool.

Aladin (see Fig. 7) is an interactive software sky atlas allowing the user to visualize digitized astronomical images, superimpose entries from astronomical catalogs or databases, and interactively access related data and information from the SIMBAD database, the VizieR service and other archives for all known sources in the field. Created in 1999, Aladin has become a widely-used Virtual Observatory portal capable of addressing challenges, such as locating data of interest, accessing and exploring distributed datasets, and visualizing multi-wavelength data.

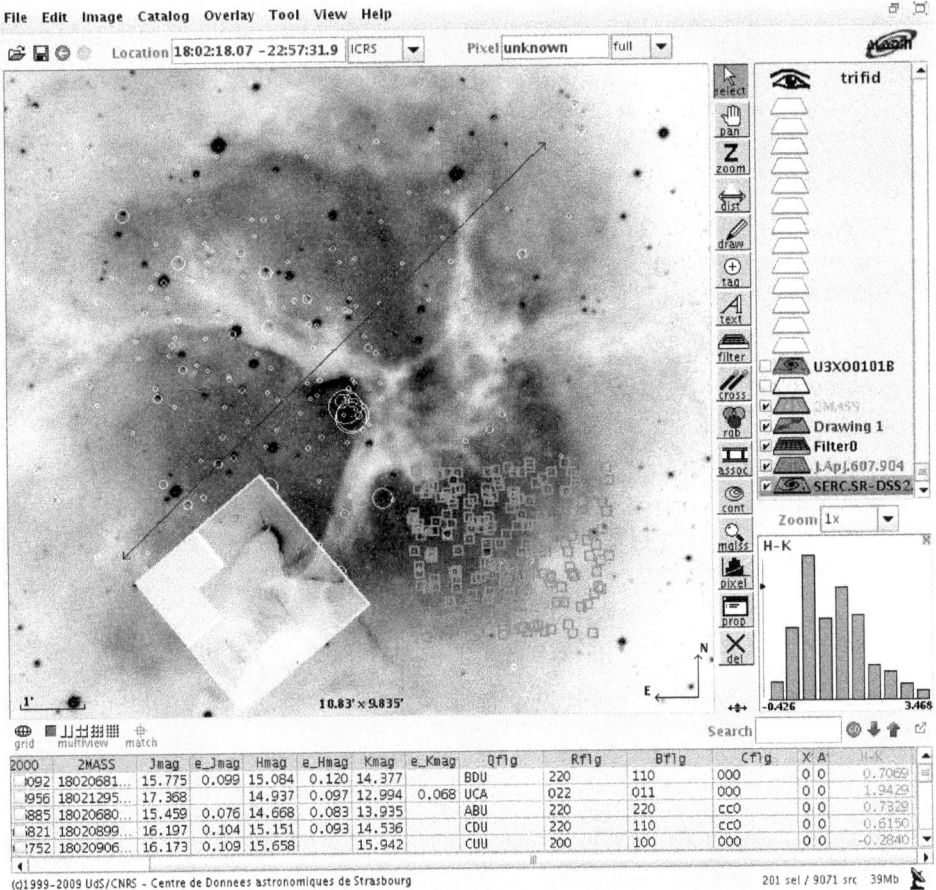

Fig. 7. Aladin: optical and HST images of the Trifid nebula. White circles are Chandra X-ray sources (Rho *et al.* 2004) with radius proportional to counts; squares are selected 2MASS sources for which an histogram of the $H - K_s$ color is displayed in the lower-right hand corner.

Aladin can be launched as an applet running in a web browser or as a standalone application. Installation instructions and documentation are available at http://aladin.u-strasbg.fr/

5.1 Accessing Data from Aladin

Aladin gives access to a wide range of data services that delivers images or tables:

- Access to images: the Aladin software can access and query the Aladin image server, a dedicated database running at the CDS and providing optical images coming from the Digitized Sky Survey (DSS), as well as infrared images from 2MASS. Aladin can also load images coming from external services and archives, such as SkyView, the Sloan Digital Sky Survey (SDSS), the Multimission Archive at STScI (MAST) or the NRAO VLA Sky Survey (NVSS), or even local images over 1 GB.

For most services, comprehensive metadata including the coverage and the orientation of the image fields help the user choose data of interest.

- Access to tabular data: for a given region of the sky, Aladin can retrieve information coming from the SIMBAD and NED astronomical databases. It can also query any of the ~8000 tables with positions available in VizieR, including large surveys such as NOMAD, 2MASS, USNO-B1, observation logs from missions such as Chandra, HST or , but also tables of individual objects published in journals.

In addition to the predefined list of available servers and to the access to VO-compliant resources, the user can describe additional servers in a configuration file, assuming they are available through HTTP calls. Moreover, local data in the form of FITS images, FITS tables, or tabulated format can be loaded into Aladin and compared with data retrieved from servers.

5.2 Visualizing and Comparing Data with Aladin

Aladin offers a whole set of tools allowing easy comparison between data covering the same region on the sky.

- Superimposition: once loaded, catalog data can be superimposed on an image of the same field. The user can then zoom in and out or pan the image by dragging the mouse. When selecting a set of objects, the associated measurements are shown in a dedicated panel.

- Combination of images: images can be combined to create a color *RGB* image or an animated blink sequence. The latter proves to be very useful when looking for objects with high proper motion or for transient events.

- Multi-view: the main window can be splitted into several panels allowing easy comparison of different images of the same region. The different views can be synchronized to show the exact same part of the sky at any time.

- Cross-identification: a cross-match tool lets the user cross-identify catalog data coming from different services, using the separation between the sources.

- Visualization of catalog sources: a tool allows filtering out and modifying the way catalog data are displayed (size, color), using information stored in the loaded catalogs. It allows, for instance, stars to be displayed as circles whose radius is proportional to the value of the flux or to display their proper motions as an arrow whose direction and length reflect the actual values of the corresponding parameters.

6   Conclusion

The CDS has constantly evolved since its creation, taking advantage in particular of the development of the Web, of electronic access to publications, and of course of the technical progress in storage capacity and network bandwidth, among others. The primary goals are to construct content from many information sources, especially papers published in journals, and to make it available to the whole astronomical community.

Metadata and normalization are absolutely critical ensuring easy access to data since they make it possible to select published data relevant to a project without having to read every paper. It also allows astronomers to find appropriate data among the increasing amount of available information (data mining). Such an ambitious goal implies a continuous concern for quality, which can only be achieved with the help of authors **and** referees. Special attention is given to providing data access simultaneously to non specialists and to providing detailed information that may be of use to specialists.

The content of VizieR and SIMBAD obviously reflects the literature, so references are systematically associated to the information stored in these databases. When errors or outdated information are detected by a user, it is advised to inform CDS, using `question@simbad.u-strasbg.fr`, and to provide as many details as possible with a reference so that a correction can be applied.

CDS has been at the forefront of networking online information, in close collaboration with the journals, the ADS, NED, and observatory archives. Thanks to the Virtual Observatory framework, CDS services, hence their selection of published information, are integrated better and better with other online resources, *e.g.*, through the Aladin portal, and through all other VO-enabled tools.

Appendix

A   SIMBAD Query Modes

SIMBAD has several query modes, using different kinds of criteria and allowing different outputs, both for direct reading and use in applications. Also some Virtual Observatory enabling facilities have been implemented.

A.1   Basic Query

This is the simplest query mode. The form contains a single field that can be filled with an identifier, coordinates, or a bibcode, and it returns an astronomical object or a list of objects or the reference summary. This query mode can be added to the list of query servers in any browser's tool bar.

A.2 Simple Query

A.2.1   Query by Identifier

Any identifier understandable by SIMBAD can be submitted to the database to retrieve information known for this object. An option allows a query around the given name, also specifying the search radius. Query with partially known identifiers is possible by embedding wildcards (*=any string, ?=one char, [c1-c2]=one char from c1 to c2, ^ as first character in the brackets means one char excluding those specified). It is also possible to submit a list of identifiers prepared in an ASCII file, one identifier per line. The result can be a simple list of all the objects or a detailed display for each object found.

A.2.2   Query by Coordinates

Queries by coordinates allow to retrieve all the objects contained in a circle defined by its center and a radius, which is 2 arcmin by default, and is limited to 10 degrees. A list of coordinates, each of which can be associated with a particular radius, can also be submitted from a user's local file. The result will contain a set of lists corresponding to each submitted position.

A.2.3   Query of Bibliographic References

There are several ways to query the bibliographic references contained in SIMBAD. The basic way is to query a bibcode to get either the paper summary (title, authors, comments, ...) or the list of objects contained in the paper. The bibcode can be written using its regular syntax, with the possibility of using wildcards, or by specifying the different parts of the bibcode: year, journal abbreviation, volume, and page. Any field can be omitted. A reference summary can also be obtained by using several criteria, namely journal abbreviations, publication year range, authors, and words in the title.

Any list of references obtained by such queries can be sent to the ADS for further querying, with the benefit of all ADS facilities.

A.3 Advanced Query

A.3.1   Query by Criteria

Queries by criteria allow to create samples of astronomical objects that share several common properties. SIMBAD offers more than 230 different criteria, such as basic data fields, existence of identifiers or measurement catalogs, bibliography in a range of years, and fields in the measurements. It is also possible to select the objects in a sky region, which can be a circle, an ellipse, a rectangle, a declination zone, a box defined by great circles, or a polygon.

Most criteria can be defined by a value, a range of values, or memberships in a list of values. For spectral types and object types, thereare two possibilities: either a strict correspondence with the given value (sptype = 'F3' or otype =

'star') or with all the underlying types (sptypes = 'F3' will include F3 type with all the luminosity classes and/or peculiarities; otypes = 'star' will include all kinds of subtypes of star, *e.g.*, variable stars, pulsars, carbon stars.

Query example: get all galaxies in a 30 arcmin circle having a *B* magnitude less than 16, and belonging to either the UGC or the MCG catalog:

```
otypes = 'gal' & region(CIRCLE,12 30 +10 10, 30m) & magB < 16
        & (cat = 'UGC' | cat = 'MCG')
```

A.3.2   Query by Scripts

A SIMBAD script is made up of a set of commands defining the output, some common parameters like a default frame, and a radius for coordinate queries and any mixture of identifiers, coordinates, bibcode, and samples to be queried.

The output is always a straight ASCII file, with no embedded HTML tag. By defining its own output format (or a VOTable output with the needed fields), the user can easily ingest the SIMBAD output in its own application, maintaining a fair independence of upgrades made in the SIMBAD server software.

Example of a script, displaying, for a list of stars, an identifier (Hipparcos if it exists, HD if not, or any first one if HD is also missing), coordinates, B and V magnitudes, spectral type, and the number of references between 2000 and 2009:

```
format object f1 "%-10IDLIST(HIP,HD,1) | %-27COO(A D) |
        %FLUXLIST(B)[%4.2*(F)] | %FLUXLIST(V)[%4.2*(F)] |
        %-5SP(S) | %#BIBCODELIST(2000-2009)"
hd 1
hd 2
hd 3
```

The output of this query is

```
HIP    422 | 00 05 08.8331 +67 50 24.013 | 8.73 | 7.43 | K0    | 2
HD     2   | 00 05 05.950 +57 46 13.22   | 8.67 | 8.20 | F5    | 0
HIP    424 | 00 05 09.7582 +45 13 44.505 | 6.75 | 6.69 | A1Vn  | 7
```

Such a script can also be submitted from a user's file, which can be prepared manually or as the result of some application.

A.4   Output Options

A dedicated form allows defining many output parameters, such as the kind of output that can be an HTML page, an ASCII output, or a VOTable, a list of data to be displayed or not, selection of a preferred identifier, range of reference years, notes. Most of the output options can be defined separately for full object display and lists of object display.

The output options are kept in a cookie and are reloaded during each session run from the same workstation and the same browser.

References

Bonnarel, F., Fernique, P., Bienaymé, O., *et al.*, 2000, A&AS, 143, 33

Cayrel, R., Jung, J., & Valbousquet, A., 1974, Bull. Inf. CDS, 6, 24

Feigelson, E.D., & Lawson, W.A., 2004, ApJ, 614, 267

Feissel, M., & Mignard, F., 1998, A&A, 331, L33

Genova, F., Egret, D., Bienaymé, O., *et al.*, 2000, A&AS, 143, 1

Genova, F., 2007, in ADASS XVI, ed. Shaw R.A., Hill F. & Bell D.J., ASP Conf. Ser., 376, 145

Jung, J., & Bischoff, M., 1971, Bull. Inf. CDS, 2, 8

Jung, J., & Ochsenbein, F., 1971, Bull. Inf. CDS, 3, 6

Lortet, M.-P., Borde, S., & Ochsenbein, F., 1994, A&AS, 107, 193

Lynga, G., & Westerlund, B. E., 1963, MNRAS, 127, 31L

Ochsenbein, F., 1994, Bull. Inf. CDS, 44, 19

Ochsenbein, F., Bauer, P., & Marcout, J., 2000, A&AS, 143, 23

Ochsenbein, F., & Dubois, P., 1994, in Astronomy from Large Databases, II, ed. Heck A. & Murtagh F., 405

Ortiz, P.F., Ochsenbein, F., Wicenec, A., Albrecht, M., 1999, ASPC, 172, 3790

Rho, J., Ramirez, S.V., Corcoran, M.F., *et al.*, 2004, ApJ, 607, 904

Wenger, M., Ochsenbein, F., Egret, D., *et al.*, 2000, A&AS, 143, 9